Man In Time

Jean Guitton

MAN

IN

TIME

Translated by Adrienne Foulke

UNIVERSITY OF NOTRE DAME PRESS

Notre Dame & London

Library of Congress Catalog Number: 66-14631

Copyright © 1966 by
University of Notre Dame Press
Notre Dame, Indiana

Originally published 1961 as
Justification du Temps by
the Presses Universitaires de
France, Paris.

Manufactured in the United States of America

Contents

Introduction
7

On Eternity
11

On Contamination
27

On Dissociation
47

On the Structure of Time
81

Conclusion
125

Introduction

EARLY PHILOSOPHY addressed itself to the task of defining the relationship of being to nature and to thought. This done, it then seemed a simple matter to discover what man was. And once man had been assigned his appropriate place in the architecture of the Whole, the problem of ethics was also resolved. In very general terms, these are the ideas that inspired the Greeks. Later thinkers found it useful to set metaphysics aside entirely, the better to discuss man. In periods when philosophical systems seemed sterile, thought descended from heaven, so to speak, to hover over the question, What is the nature of man? In the West, humanism has coincided with speculative skepticism. Kant, Hume, and Montaigne come immediately to mind.

7

Must we, then, choose between two inevitable courses, one of which leads us to explore the absolute, while the other confines us to the study of man? It seems to me that, on the contrary, they must somehow be harmonized, that this is indeed the price of true balance. A philosophy of man that offered no insights into the composition of his essence would be merely descriptive, or it would be even less—a simple reconstruction. Witness the empiricists, who believe that they are supplying an explanation, whereas actually what they offer is mere narration or recombination. A philosophy of being that sought to deduce from some essential substance what human nature is, would make human nature incomprehensible. At least one philosopher did attempt to conceive of man at the center of being without ever deigning to speak the language of conscience—I am referring to Spinoza, of course. He is illuminating, as only a sublime failure can be. One could imitate his example, but one could not repeat it.

However, is there not a middle way that belongs to an intimate, profound tradition? Following this method, we would apply ourselves to the study of man's inner nature; we would try to discern the structure of his being in his various states of consciousness and, by analogy, the structure of all material nature; it would then be easy to recognize wisdom, for wisdom would be voluntary submission to the discipline of essence. Is this not what Plato attempted in his later life? In the *Philebus,* he deals with morality but, by way of preamble, explains the most general forms of being and the modes of their combining. This was also Pascal's idea in *Conversation with M. de Saci,* where man is studied in terms of two dimensions, abjectness and greatness, leading Pascal to recognize the laws whereby, without confusion or separation or even equivalence, opposites are brought into a balanced unity.

Mindful of these great examples, I should like to examine certain aspects of the inner nature of man and, in this intimate realm, explore what temporal being is in relation to eternal being. The scope of my inquiry will be extremely limited, it is true. What follows here is only part of a more extended treatise on human nature. Were someone to object that it is useless to try to give an idea of the whole by detaching one of its parts, may I say that such an attempt is at least excusable when the part in question is not so much a fragment as it is an image or monad in which the form of the whole is reflected.

To avoid any misunderstanding, let me also point out at once that time in the sense in which I shall be discussing it here is not the time of the mathematician or the physicist. It is not the pure, profound duration that consciousness apprehends only through some special effort or extraordinary experience. Time as I envisage it here is the time of human life, the time we learn to know as it flows by on a quite ordinary level. Unfortunately, neither of our two means of knowledge and expression—philosophy and art—has often been concerned with the ordinary man and with ordinary life. So often the ethics that evolve from philosophy seem suitable only to the minority, to whom salvation is thus restricted. And classical tragedy, like the modern novel, shows us primarily the aberrant aspects of human existence. This is not to say that the most abstract metaphysics or the most awesome tragedy cannot teach us something about our voyage through life. We are indeed better able to know ourselves as men thanks to the *Ethics* or to *King Lear*. But the ordinary conditions of life are those that escape us most completely. One needs much experience to relish the routine.

I have purposely not quoted other contemporary authors, and I have not referred to anything I have written on analagous subjects. There is probably not a single line in this book for which I am not indebted to the influence of others, but I would rather that this be made clear not by references or acknowledgments but by the interconnection among the book's several parts.

On Eternity

THE PROBLEM OF TIME—I mean time both as it is lived by man and in its relation to eternity—is unfailingly to be found at the core of all metaphysical and ethical questions in a form that is at once substantial and fleeting, as if, wherever it lodges, it sparks a dilemma made in its own image.

Certain states of consciousness that we may apprehend via either feeling or contemplation seem to contain some element of eternity. What do they mean? What value have they? For example, what is happiness? Does the ideal that we pursue in life endure beyond time? But then, what is time? And especially, what is the present, what is this moment that is forever escaping from us and is yet the only one in which we exist and have

11

our being? What is the relationship of the present to the past, which is no longer, and to the future, which is not yet? How can we remedy time's passing, check its flow, make it *be* forever, and so satisfy what is perhaps the most fundamental desire of mankind? Shall we look for eternity in the depths of the present or advance it to an aftertime?

Let us try now to fix our attention on this mobile point that is our present. No experiment lies more readily at hand, for the present is offered to us anew with every passing moment. Yet none is more difficult, for nothing eludes us more than does the present; if ever we were able to fathom it, we would surely learn one fundamental aspect of being.

Anyone who sets out to analyze the present passes through three kinds of knowledge. First, letting himself be guided by a general assumption that coincides with his own wishes, he believes the present is stable, solid, and substantial. Is it not the only part of time in which we are capable of sensory experience? If thoughts are in conflict in the present and transformed into pain, are not our pleasures more luxuriant and our joys prolonged? Furthermore, is the present not, as it were, the act of extension, the one plane on which our bodies exist and have consistency? We are tempted to suppose that our unconscious self is removed from the present, because it eludes our immediate apprehension, but nonetheless the present is richer than that which is merely current. Where shall we locate inclination, if not in the present, that is, in the moment that recapitulates the past and prefigures the future?

However, if we wish to define the present more exactly and seize it as it passes, we are immediately in difficulties. The present is constantly disintegrating into two parts, both of which have the specific characteristic of being not present. The first part is made up of what has just been and of what is ceasing to be. The second, which is the principal element and the one that gives to the present its momentum and its form, is a surging of the human spirit toward some potential, toward some point established by will or by desire, from which action ensues: it is an invention of what will be and, at the same time, a passive waiting for that which is about to appear. We tend to conceive of these two modes of the present as places. The mind envisages a domain where memories are preserved, which is symbolized in the convolutions of the brain; the mind likewise envisages futurity as a locale where the present is awaited or, rather, into which the present penetrates. Prophets are able to foresee snatches of the future and scientists would perhaps be able to calculate it if only they knew the past of the universe in its entirety. But we need only think about what "place" means to be convinced that neither the future nor even the past exists in any place, as things do. Rather, they are two aspects of being (like two slopes of a mountain) that divide an action in progress. The content of this action is continuously dissipated; either it is not yet, or it is no longer. Once we have experienced this flight, we may well ask ourselves whether the phrase "to be" can still have any meaning for the thinking man.

And yet, if we reflect still further, and especially if we have lived more in terms of the spirit than of the senses or the intellect, the analysis that dissolves the

13

present may be deemed irreproachable by the intelligence but will not seem to correspond to a profound truth. I can indeed dissipate my substance, but I am aware of this dissipation, which proves that it is not total. Habit enriches my present with everything that flows into it from my past existence. Through memory I find the past again, and it seems to me that if all my efforts at recall were to converge, I would restore the past and establish it in one immobile instant. No parallel effort to summon up the future is possible, for by its nature the future is unforeseeable. I cannot ascertain future events or divine the acts of other men or even be sure what my own conduct will be. But if I think instead in terms of my private inclinations, then I do feel that I possess obscurely the substance of what I am seeking. The immanence of time is more readily perceived as physical vigor is impaired, that is, as the body ceases being the threshold to the soul and becomes instead its support.

The man who wishes to know himself, however, does not wait to reach old age: he tests his identity, he experiments with this profound life of the self that escapes time. Everything that takes place happens within me, it is true; it is equally and perhaps more true that what happens causes me to grow in dignity and that time is a nourishment for which even thought hungers. Without this nourishment, thought would still know its own truth but would be deprived of its density and full maturity. I have said that we flow in time. It is probably more true that time flows in us.

What we have observed in this analysis of a moment of present time we will encounter again in every domain of the spirit. At first glance succession and permanence appear to be in opposition, and we are forced to oscillate between them. But substantially they are united.

Whenever some external reality is modified, we unfold its successive states over a period of time. If what we are dealing with is a state of consciousness, it will seem even more subject to time than change of a physical nature. Consciousness and duration seem to be so much a piece that we are tempted to define them in terms of each other. But if time is, as it were, the form of sentient existence and even of spiritual life, eternity is the form of thought. In every exercise of the intelligence we detect an attempt to suspend and even to suppress time. The early philosophers saw this clearly—perhaps all too clearly—since they often confused the intelligible with the eternal; this obliged them to relegate the individual feeling, moving, enduring element in human beings to nothingness and to conceive of eternal life as a logical life absorbed in the contemplation of unity. Aristotle maintained that what can be mentally conceived is always the same and always complete; for him, the intelligible, the eternal, and the general were associated. Because pure action was eternal, Aristotle could forgive God for being unique of His kind.

The system was oversimple, of course; it had to be made more supple, and the task of logic was to restore intelligibility to the contingent, the singular, and the historical. In a parallel way, philosophers have been led to invest eternity with duration and to distinguish in it

15

actions, progression, and phases. But no one can deny that there is a kind of affinity, a kind of resemblance, between the substantial and the eternal, as there is between the accidental and the temporal. The intelligence is always searching for relationships that are not subject to change. Even when the mind considers the question of becoming, it tries to discern either substances and forms or laws and types. The existence of plant and animal species inclines us to condemn time, which seems to affect only the individual. As Buffon said, a continual succession of consistently similar beings is equivalent to the perpetual existence of a single one of those beings. And when the mind turns to recomposing the historical past, it does so in order to discover analogies and rhythms; if it becomes interested in a unique fact or a single personality, it assigns that fact or personality to a type that constitutes a species all its own.

However, the temporal and intemporal elements we have just differentiated are not set in juxtaposition. Much less are they fused. Their relationship is that of form to matter, and this may be observed sensorially. To perceive a red light is to be able to gather together in one indivisible instant (a true image of eternity) the four hundred trillion distinct vibrations that, were they to appear one by one to our visual sense, would keep us busy "seeing" them for twenty-five thousand years. How could we unite them and perceive them without the action of a mind that is contemporaneous with each vibration, as divine eternity is contemporaneous with each moment of history?

It is possible to conceive of spiritual beings who would live in time as do we, but whose memory would

16

be very different from ours. These people would be capable of being present again in their past, that is, of reliving this or that past moment as they chose. Our human memory never recaptures the present from our past. We struggle to make up for this: the literary forms of novel and autobiography have been brought to maximum scope and effectiveness in order to create the illusion that we can buy time back. But when an author succeeds in resuscitating a remnant of his past, he seems to have done so by accident, almost furtively. Why does he fail? Must we suppose that we are in such bondage to current action that it hinders retrospection? Or is it because memory is alien to the sense of touch, whereas it is responsive to sight, which of all our senses is the most neutral? Might it not also be that even when memory is concerned with specific things, it is much nearer to reason than we generally assume? In any event, memory brings the past to a standstill; it spreads the past out like a canvas and presents it to us all at once, so that we are able to repossess the past through thought exactly as we survey space at a glance.

Given these conditions, memory never recovers the past as it was: the recollection of each moment is impregnated with the confused memory of all that preceded it and especially of what followed it. This is enough to make the past-in-present quite unlike that once-upon-a-time present whose future was then inexplorable. Memory transforms its material; each of us is the artist and, in a way, the prophet of his own past. We discern resemblances among events, affinities between people, and the mighty concert of chance. We recompose our own histories, arranging them to lead up to

17

the present moment that we account for in terms of these histories. We color them with opinion.

Many people imagine that they prefer the excitement of the present to the calm of a past that is forever settled. Who could say they are mistaken, for remembrance is a disconsolate thing; in memory, action is no longer possible, nor is pleasure, except by artifice, and possession is denied the senses. Nonetheless, sensory experience would be worthless if there were not this possibility of its reappearing in the intelligible light of memory. As intelligence develops, man relies less and less on sensation, and more and more on memory. It would probably not be too difficult to identify a long line of rememberers among spiritual men—people who discover something only by recovering it, who have the impression that what they now love they have loved before, for whom meeting is recognition; such people consent to move forward toward tomorrow only because tomorrow will become yesterday; they resign themselves to seeing only because that is the prerequisite for having seen. Thus it happens that there is much of the timeless in the act of remembering.

If we now carry our analysis forward to the area of the will, we can say that time is the locale of options, eternity that of destinies. In the course of a temporal life, we always have the opportunity for commitment and perhaps even greater opportunities for disengagement. The most secure marriage can dissolve because of mere lassitude, and if fidelity has merit, it is because fidelity is a continual victory over inconstancy. Since man's propensity for giving himself in a moment of enthusiasm is as great as his perseverance is minimal, an ill-placed love is somewhat redeemed by the mere fact that it lasts. So long as a man lives and is master of himself, that long is it possible for him to modify the direction of his good or bad intentions.

Descartes seems to speculate whether there does not exist in each of us a certain capacity for love that will develop into virtue or vice depending on our upbringing and circumstances, as well as on an initial endowment, which is the quality of our spirit. Fortunately, it is rare and perhaps impossible for virtue to deny itself in the end, but one does see noble natures resort to evildoing throughout a lifetime and redeem themselves only at the last. For all of us, the arrival of death means that the future is blocked. Nothing more can occur, no further events can take place, and our existence becomes a work of art for the spirit to contemplate. Even if the soul had no future once it was abstracted from materiality, our death forces others to conceive of us in thought and so gives us an unalterable form in their consciousness: our death changes us, both in them and in ourselves. But this empty, cerebral eternity is only the symbol of the real eternity where we are transformed into the being that is the creation of our deepest will. It is this immutability of the eternal being even more than its spirituality that makes death a cruel thing to think of; men find it hard to imagine life that no longer involves a quest. Therefore, some believe that the adventure is pursued after death. Yet if we examine metempsychosis and spiritism, we find that we are not dealing with a new existence but rather with the prolongation of present life, which is the only life where

the drama has really been played out. Thus the concept
of eternity almost implies necessity. And let me point
out in passing that this explains why we have difficulty
in conceiving that God is free. How, in an immutable
reality, can there be room for option?

We would be wrong to envisage the will's eternity as
absolutely contrary to the temporal life. The eterniza-
tion of our being is manifesting itself as of this very
moment, and we could say that at any present moment
the will is supported by a timeless element. This notion
is implicit in the popular idea of destiny, which we will
want presently to examine.

How hard it is to recognize the true events of the
recent past, and to identify true beginnings! Nothing is
more obscure than contemporary history to the person
who wishes to distinguish the essential from the contingent. As we move further back along the highway of
time, we are fairly able to descry the crossroads where
history wavered among several equally possible futures,
but we still remain in doubt about dates of origin.
Obviously, the moment when a war broke out is verifi-
able; on closer scrutiny, however, we perceive that this
beginning was an ending. Let us suppose that presently
we manage to define the outmost boundary of this pre-
history. We will still feel that, with a little more per-
spicacity, a little more knowledge, we could identify
this "first cause" as a consequence; in all events, if absolute beginnings do exist—the existence of heroes, for
example, would prove as much—we are quite incapable
of determining where to posit them. This is why it is
difficult to establish responsibility. The history of a temporal action is divided into two phases separated by a

20

period of unawareness. The first phase of the action commits us or, at the very least, lays the groundwork of habit, and if it represents a failure of will, freedom has already been diminished thereby. But in this first moment, the individual believes he is released from the chain of cause and effect; he has the impression that he is embarked on an experience that has no boundaries. At the moment we commit ourselves we do not recognize that we are doing any such thing; to us, our choices seem to be trials or adventures, actions as gratuitous as the moves we would make in a game. However, the day comes when we realize that we determined our destiny at the very moment in the past when it had seemed to us we were merely juggling with possibilities. If our present is auspicious and legitimate, we feel that a helpful Providence has stimulated and guided our past choices, but if our action is reprehensible, it is already too late and from this we sense our errors as being irreparable. We should like to possess liberty here and now, we search for it everywhere, and are astonished not to find it. The reason is that freedom and time are not interchangeable. Some men have thought that if we concede the possibility of there being a single free act in the course of a lifetime, it is enough to prove that freedom is not simply a word; logically, indeed, a single action does suffice. But we will be nearer the right if we post freedom at every step along the road of life rather than smother it everywhere in order to rekindle it on a single issue. Freedom belongs to every moment of life and cannot be confined to any one moment. Freedom resides in the act of resolving, and resolution is an inner action; it exists in the act of willing, which is a continuing action, rather than in the act of choosing, which is rare. Furthermore, even when the will is limited to an act of consent, is it not a constant choice? Freedom is an act of intent that derives from the quality of a human being rather than from a capacity for decision that manifests itself only in crises.

The man who makes a given decision often feels as if he is merely assenting to a resolve that has already been taken; any resistance to it would be drawn from a part of himself that may still be powerful but is no longer or, at least, is less and less the true "I." To quote St. Augustine's nontranslatable phrase, *"Ibi magis iam non ego."* The same phenomenon is to be observed in the awakening of a great love: when love dawns on the conscious mind, it appears in the guise of an earlier habit of the heart, whence the awkward language, the abundant silences, the plethora of symbols. The spirit, not the body, seems to have moved first, which may account for the timidities. Thus, our actions anticipate even more than they follow us. The strange thing is that with a free action being divided into two, as it were, we yet do not have the impression of willing something so much as of consenting to something already willed.

The psychologists have studied this sensation of *déjà vu;* its unhealthy forms must not make us forget that the same phenomenon can occur in normal perception, on condition that interest and attention have flagged. But there is also an "already willed" and an "already loved." Experiences of this order support our notion of destiny, with its accompanying train of concepts and images: our childhood, our role in life, good luck and bad, and the idea that there are people who are made for each other, people who will meet and recognize each other. These are imperfect formulations, for they allow us to believe that destiny is decreed in advance and can do no more than unfold; yet, error for error, such language is more respectful of the individual than the terminology used by the partisans of necessity, which makes man out to be one of nature's products, or explains existence in terms of a primitive choice the memory of which is lost to us, or confuses freedom with impulse.

The formulations of destiny may be inadequate, but each possesses its own symbolic truth. The defect they share is this: they lead us to believe that they propose a solution, whereas they merely register the presence of a nontemporal element that penetrates but does not adulterate human time. Kant maintained that this element is operative in a timeless instant. It was an awkward concept intended to translate this anteriorness of what we are to what we are, which signals the fact that the substance of time eludes succession.

The last role in which time and eternity are opposed concerns communication between people in love. Time has always been deemed the realm of elusiveness and of symbol, eternity that of presence and of transparency. People are separated by their bodies and are able to communicate only by language. Since visual images are too ambiguous to communicate a thought and since, furthermore, the slightest screen is enough to destroy their effect, we have had to choose the mediation of sounds rather than signs. Spoken language is subject to time, however. Despite all our efforts to make up for the law of succession (of which poetry is the most perfect and syntax the most useful example), we can attain to truth only by revision, reservation, and repetition. By way of contrast and compensation, we dream of an eternity where all distance, all absence, is abolished. This eternity, which is simultaneous and entire, frees us from language, which is successive. Physical bodies having been eliminated or sublimated, one consciousness becomes accessible to another; we might say that they are all "word."

In order to conceive of the society they then constitute, we must think in terms of groups where people communicate not so much via conversation among themselves as by participation in a commonly held belief. It is quite impossible to imagine the metamorphosis whereby a temporal consciousness would be transformed into an eternal consciousness, but it is less difficult to think of a temporal society's being transformed into an eternal community in which love is the bond. In this case, nothing essential has to be modified; we need only discard our mortal bodies and language. This is perhaps why several religions conceived of salvation as collective before they narrowed it to individuals.

The presence of eternity in time is not to be excluded, however; it has already commenced in the form of an intimacy observable in the higher forms of affective life. Every emotion we can distinguish in the obscure, potent unity that constitutes our fundamental human drive tends either to unite our being with the object that alone appears capable of completing it or, on the contrary, to divert it from whatever threatens to destroy it. For example, aspiration, which together with love—and perhaps more than love—is the basis of affection, is an attempt to reach out toward the desired object by shrinking the duration that separates us from it. But let us think for a moment of the final or definitive state, namely, possession of the beloved object. For all its intimacy, this possession would not be fusion; the good we aspire to must remain distinct enough in order to have between it and us that minimum distance without which affection itself would be snuffed out. Love eagerly seeks the words that express identity, yet it finds such abstract unity so colorless that it has recourse to more fervent terms: it finds "union" less desirable than "fusion." However, love would be caught in its own trap if it were to achieve perfect consummation; distance is the mediator of unity, and absence binds individuals together whom the community separates. This intimacy, which may be putative or actually savored, but which is always precarious and compromised, forms the essence of many emotions. And so closely connected is it with eternity that we are shocked when intimacy ceases; it is as if eternity had been rent asunder.

Converts, especially those who have taken the step after long hesitation, would experience some such feeling as this. Here again—and this may be the perceptible root of the idea of grace—the convert feels as if he were approaching a world that existed before him, that he is catching up with a being who has been waiting for him or, rather, who has loved him first. He cannot understand how this love could have commenced in time, so he interprets whatever period that preceded the date of his discovery of love as a long preparation. The most celebrated example in this connection is St. Augustine, but his example would not be so famous if many people did not believe they recognized themselves in him —notably Pascal, who has the liberator say, "You would not look for me if you had not already found me." "Already!" The word is charged with meaning. It indicates that certain states of consciousness cannot be lodged in the present only, for, immediately as we begin to consider them, our train of thought is led back into the past in search of a starting point; this we never firmly establish, and sometimes we must carry the point of origin back to a realm so remote that "already" is almost equivalent to "always."

Thus, in order both to prepare for balanced action and to respect the structure of our being, we must explain that eternity is beyond time but that nevertheless we already exist in eternity. Or the problem could be put this way: we must respect the virtual eternity of the present without excluding the true eternity that we perceive in the guise of the future, and recognize that eternity can be defined only by abolishing all attributes of succession. This presents considerable difficulty. How are we to maintain at one and the same time that eternity in the present is realized within the limits of an ephemeral human life and yet is removed from our experience? How can we counsel man to establish himself in life and at the same time to disengage himself, to savor his present life in all its fullness and yet desire to escape it, living in time as if he were not of it? Anyone who perceives how we can reconcile these contrary demands would not be far from knowing what is man.

On Contamination *

We have seen how human duration is divided into two inseparable yet distinct elements. One functions so that the modes of life are not concentrated in any unity but follow one upon the other, and constitute time. The other element mends that which was about to come undone; it reunites that which was being dissipated; it provides links for duration. It is a profound continuity, a virtual presence, an unceasing aspiration. The two factors interpenetrate, they are interwoven— or let us say that each implies the other, each requires the other as do form and matter.

* In recent philosophical usage "contamination" refers to the reciprocal influence between two or more notions in contact leading to confusion between these notions.

27

However, the inner time that they together compose is a kind of substance over which freedom has influence. I am not forced to submit; I can intervene in time by thought or action—or better, I can intervene by an act of thinking that will modify the composition of these elements. This possibility of acting upon the very action which is my own life is an attribute peculiar to humans.

How will I put it to use? My first move will be to try to link the eternal and the temporal so closely that any contradistinction between them should disappear.

If only we could make the temporal eternal! Actually, this would not be so much a matter of transcending as of suspending time, or rather, of filling it so full that it could no longer slip away. But where can we look for help in this? Do we as human beings possess some capacity that would permit us to halt time without having to slip outside it, simply by increasing its density to the absolute degree? We should have to search for such a faculty in the ambiguous zones of consciousness where the biological and the spiritual are sutured; here we would encounter forms of thought that transpose the impulses of animal nature and so prolong them. For lack of a better word, I will call this faculty "life." I will say that the genius of life consists in contamination, and that we go along with the resulting confusions all the more gladly because our intelligence is thereby spared the effort of true discernment and our will is not asked to submit to the travail of sacrifice.

There is one tempting way to think of the future, the past, and the present, the first and most immediate effect

28

of which is to adapt us to an even, comfortable way of
life. If we will imagine the past to have been better
than it was, believe that the future will be palpably perfect, and inflate the present with voluptuousness, passion,
or mere infatuation, then we will have carved ourselves
a little niche in the universe where we can thrive. These
attitudes of the soul (they are not accounted for by their
psychic mechanisms alone) correspond, although they
are of another order, to the ways in which animal instinct functions: they help us adjust to our surroundings;
they prevent our pining and wasting away; they anesthe-
tize us to the fact of imminent death by forcing us to
function in the present or in the image of the present.
Life construed in this way makes us move and have our
being in a contaminated milieu that could be defined
equally well as temporal eternity or eternalized time.

A critical intelligence would make short shrift of this
illusion, of course. And it is very beneficial to expose
these amalgams wherever they occur, for they misdirect
the motions of life and are not so much thoughts as
pressures. The best way would be to begin with the
simplest and most readily known adulteration, but it
happens that the most common forms of contamination
concern our perception of time. They are difficult to
intercept and correct because they are as inextricably
a part of self-awareness as the distortions of perspective
are a part of vision. The adult common sense that corrects visual illusion is paralyzed, however, when it
confronts the illusions that affect subjective time.

We will first examine how we perceive the future, for

contamination is more noticeable here than anywhere else.

The future is the locus of illusion. The observation is so banal that one feels almost embarrassed to repeat it. Yet having said this, have we gone on to explore the ultimate reason why the future is illusory? It is not only that by definition the future is not yet given, and that no matter how sure we suppose it to be, it is always improbable; early experience inculcates in all men a beneficial insecurity which helps them correct their certainties. It is not even that the future as we prefigure it does not satisfy our expectations on all points, for in this regard illusion affects only details. If we wanted to indict our customary image of the future, we would do well to remember that it conceals death from us, for until our last moment we anticipate yet another moment, and with our last sigh we count on drawing a fresh breath. But we do know from experience that other people die, and this helps us to correct illusions about the future's enduring forever. Errors such as these can be corrected, of course, but side by side with them there exists, I believe, a fundamental illusion. The real illusion we entertain with regard to the future lies in our conviction that the moment the object of our prayers will be granted us, then, by some mysterious operation, time will be brought to a halt; as if time's nature will somehow have been changed, we think it will be immobilized.

There is a surplus of life that finds no place in the limited confines of the present, and we are not wrong to project this into the future. Nor are we mistaken to think that there is some essential relationship between eternity and perfection. To speak of God is to assume a relationship of this order between the two. How could one worship a Creator who did not have some right to exist eternally? And from what could He derive His title to necessary existence if not from His divine perfection? Furthermore, we are not always wrong to insist that the future will be better than the present; even as we concede that it may prove to be worse, our skill in turning misfortunes to good account may develop proportionately. The fallacy is to imagine this "after" in the form of a "forever." Let us say that the illusion is to construe the future as "eternal temporality." If we are not careful, the future that looms on the horizon of all our present moments will be a future without futurity; it will be merely a projection of the present but without flux, without expectation, without anxiety. Because this mendacious image undervalues the present, it prevents our dwelling in the here and now.

Need I say that the same illusion obtains in relation to the past, and that we visualize the past in a rosy light? Our perception of the past is highly colored by illusion. I have shown how memory erases the flow of time, how it eliminates every element of risk, and how all of the past becomes thought. But we do know that the flow of time has never been suspended; any illusion we might spawn as to its having been halted in the past would be exploded if we were confronted by the evidence of its uninterrupted succession. I might also point out that our representation of the past is far more intelligent than our image of the future. Everyone projects the deepest and most confused parts of his being into the future, whereas in evoking the past we are disciplined by reality.

Everyone's spontaneous image of the future is a mixture of boundless aspirations and infantile dreams in which that vital and heterogeneous thing we call mentality is reflected. For this reason, neither too attentive nor overprolonged speculation about the future is instructive. The corrective experience life provides is lacking, and without it the mind risks accepting as real what is actually the product of a chimera or the sublimation of obscure areas of its own. I would be prepared to put it thus: the future is mental, whereas the past is spiritual. In any case, when illusion affects the past it may color the incidental, but it will never attack the essential. With the future it is quite otherwise; here essence is falsified.

Illusion is the more observable when it is reflected in the magnifying mirror of social aspirations. When men envisage the perfect political state, they cannot conceive of it as being anything other than eternal. This is strikingly evident in the history of the Jews, who in this as in other realms have spelled out for us in bold type the drama of man's soul. The Jews dreamed of a Messiah who would lead them into an era of triumph. Yet how could the coming of this secular king fail to unleash their destruction? The advent of the Messiah will bring about one of two things. Either, one will enter upon another order of duration via a reversal that will be analagous to the one described in the myth in the *Politics*: if old men are not to become young again, at least flesh-eating beasts will turn once more to the eating of grasses, rivers will flow the year long, and women will bring forth children with joy. Or, time will stand still and the kingdom will endure forever. For a long time the Jewish spirit was understandably torn between these two concepts. But whether the Jews awaited a flesh-and-blood Messiah or a leader for all eternity, it was conceded that in neither case could time as we know and live in it continue once the Messiah had come. There was general astonishment that the early Christians, who believed that the Messiah had indeed come, should have discounted His immediate return. What would have been truly strange, however, was if they had felt otherwise, for the appearance of excellence makes it mandatory that the flow of time be abolished.

Men have witnessed the precariousness of empires, and that experience should have preserved them, or so it would seem, from these illusions. History should have made us understand that a state which pretends to be at once temporal and eternal is an absurdity. But however powerful we assume the lessons of experience to be, they have no influence over vital illusions: we see death daily, yet we do not live as creatures who must some day die. When illusion is wedded to the individual's very structure, when it is undeniably necessary to his equilibrium and functioning, what imaginable test could controvert it? Think of Christians themselves, who have had so many reasons to establish themselves conceptually in the true eternity. We see how often through the course of history they have dreamed of eternity as temporal or, again, relied upon imperial organization. (It was Eusebius who rediscovered "the kingdom of the Saints" in the empire of Constantine.) The best of them have needed a vigorous hope and faith if they were to elude the blandishments of the life impulse. Where these essential checks have been lacking—for example, in the Socialist systems of the nineteenth century—we find the Messianic theme allied with the notion that time can be halted at last; the concatenation of history will terminate the moment justice and perfection are achieved.

If we stopped to think how absolutely improbable such stability in perfection is, we would be brought, by an irresistible reversal, to assign the attribute of eternity not to the moment of perfection but to the process of perfecting. We would raise history and historical development to the dignity of primary reality. Yet if this doctrine were intellectually accepted, it would be tenable only under difficulties, for, by annulling the idea of truth, it would, like all forms of skepticism, risk running away with itself. A process of becoming that has no aim other than a continual becoming, a truth that is destroyed by virtue of endlessly surpassing itself —are such tenets thinkable? This is why the philosophers who have assimilated truth with its history, and who have conceived of this history as limitless and truly divine, have never held with the notion of unstable truth; they have preserved the traditional ideal of truth as firm and immutable. We see how they shuttled from the one notion to the other: now they canonized the whole movement of history, and again they judged this or that particular moment of the present or future to be invariable. What more typical example of this than Hegel? More than any other philosopher, Hegel has represented truth as ever unfinished, self-engendering history; he sees no possibility of assigning a terminal point to this movement, which is the life of the Mind. Yet at other times, he considers that this progress has produced his own works, that his philosophy is the goal toward which the striving of centuries has aimed —or better yet, that the eternal is henceforth represented by the Prussian state. The Prussian state alone would seem to be capable of stopping time. We find a kindred contradiction in Marx, who posits the necessity of movement in history but does not perceive that, after the revolution, evolution is still possible.

If contamination reigns over the future, it can affect the past also; generally it does so in the phase of life we call old age. The moment comes when our sense of the future grows dimmer, and it seems that then by way of some compensation the mind transfers its notion of eternity from the future to the past. The original Golden Age was not invented by happy men. A Golden Age makes its appearance when the expansion that is natural to life encounters some resistance; either it collides with some obstacle or it wears itself out, for nostalgia, which is expectation reversed, can only be born from the future's having been stayed.

At that point, the contamination will consist in choosing from the past one period that will be set up as a norm in terms of which all else will be judged, the other moments of the past being neglected. It is strange to observe how in the life of elderly people the past period that lights up with fresh life is often one that has been most fraught with danger. However, a danger survived is a false danger, for it has lost the element essential to any present danger, that is, the risk that has to be run. When in the same recollection we observe that we both gambled and won, we are left with a feeling of joy and triumph. Memory is not designed to remember hazards, and that is why gamblers lapse into rashness; a moment comes when the recollection of their past feats is purged of any detail that suggests danger, and they picture only a series of reasonable successes.

But let us go back to the macrocosm, to social life. Here political thinkers abundantly exemplify the arresting of time. For example, we find them assuming that the ideal society was achieved at some definite moment, in some early Atlantis, in a period when a given society was founded and perfected; that society, they say, must be restored or reproduced. We need not push far back into the past to find such an ideal period. It can be quite near to us in time, provided that between it and us there has been an interval during which contrary conditions have prevailed. Thus in France under Louis XVIII the people restored a very recent "ideal" past; the intervening revolution had sufficed to rehabilitate it as something near to perfection.

Restorations of this kind are much less faithful than people claim them to be, because the new forms share in what has taken place betweentimes. However, we are easily persuaded to believe that it is *the* past that returns and is preserved. Actually, it is not *the* past but this or that fragment carved out of the past and elevated to appear as the whole in a process quite as artificial as that of the modernists who canonize the present—or, to be more precise, project the present into the future. What beclouds the issue of tradition is our readiness to believe that faithfulness to the past consists in maintaining against all obstacles the forms, usages, and laws of a bygone period; actually, we should be concerned to rediscover the spirit of a past period and to create for it new molds and translations that are scaled to the present. When this is not done, the Church declines and the State is corrupted.

There is, finally, one last form of contamination; it affects the present, and I shall call it by a very old name —pleasure.

Presently I shall come back to discuss the attitude toward pleasure that I believe characterizes the rational man. Here let me say simply that "to enjoy" is to exhaust the resources of the present while simultaneously freighting it with a fictitious density. There is no pleasure without prior appetite; it is this avidity that, passing from the subject into the image of the object desired, gives that object its depth. We can derive pleasure through the senses, for each of our senses is susceptible of knowing the thirsts that communicate a new dimension to the object sensed; if human beings have known pleasure especially through the sense of touch, it is perhaps because touch, more than all the other senses, conveys to our mind the illusion of penetrating to the innermost intimacy of another being. But we can also know pleasure through the mind alone when, becoming dissatisfied merely to contemplate and love a spiritual essence, we try to appropriate it. Lastly, we can take pleasure in ourselves. In all these instances, pleasure is a seizure, a sucking in of essence, that exalts both our animal and spiritual powers. In all these instances, it also enfeebles some root in us, and in this respect pleasure is related to sacrifice. It could even be said that pleasure is absolute sacrifice, since it is a sacrifice made without reservation and without hope; the moment that follows pleasure finds the person alienated, closed in on himself, sometimes a stranger to himself. But such effects can give little cause for surprise if "to enjoy" means "to contaminate" in the way I am using this word.

To enjoy is to try to give any moment in time a kind of third, sensible dimension. Above all, it is to act in such fashion that the momentary dimension acquires, thanks to a certain gift we bestow on it, absolute fixity.

This is a deceptive attraction, of course; we can always explain it away in terms of sickness, yet were we to do this we would have to think of the sickness of a healthy man. There are people who have lost balance and who can no longer adapt to life. However, when such people are described to us and are termed "disturbed," we do not feel that they are so very remote from ourselves. Some tendency in us has become too pronounced; some attitude that exists in us only as a potential, they actually live out; but whatever the difference of intensity, we find a relationship between what happens in ourselves and what we imagine about them.

The same thing occurs when we read the biographies of great lovers. For that matter, who among us would be responsive to tragedy if great passions were not privileged to offer us an enlargement of what exists or at least could exist in each of us? If the sick, and in particular the mentally sick, did not suffer from disorders comparable to those of the healthy man, would we be

as curious as we are about asylums and mental therapy?
Would we indeed understand what the sick man means
when he moans aloud?

By way of example, let us think for a moment about
anxiety. Generally anxiety, like overexertion and obses-
sion, is ascribed to some inner burden that consciousness can no longer bear. Everything that happens in the
individual suggests a disproportion between the quantity of energy he must expend to maintain his inner
balance and the energy potential he commands at that
moment. Neurotically anxious people not only feel that
they have been inadequate but also cannot ever stop
feeling that they are being dreadfully pressured. Everything, no matter what it may be, puts an urgent demand
on them to do something. One such disturbed person
found words to describe his illness that, had they come
from someone else, would have deserved our admira-
tion: "I never feel that the given moment or the hour
I am living is really full. I must add something more
to it and then perhaps it will be better." Another
remarked that he found it "dangerous to get really
involved with an idea," because he then drove himself
to the point of exasperation. Another seemed "not to
understand what it is to be active except as an expression of anger." This man added: "I don't feel happy or
sad, I just feel as if I'm always under pressure." Here
we readily see what Leibniz meant when he spoke of
the "small, imperceptible solicitations that keep us on
the move"—impulsions that are like so many springs
straining to be released.

Leibniz had a particular genius for analysis; he saw
in this constant tension, which thrusts us outside our-

selves, the stuff of pleasure and the support required for present action. This may be, but can we fail to see that this urge may easily slip into excess? What Leibniz was inclined to conceal, Pascal vigorously denounced. In any case, this excess of action, which tends to occur in the present and to invest it with a kind of explosiveness, surely lies at the root of passion itself.

At the outset, passion contains a certain precipitancy of thought; within we are all turmoil. As the author of *Discours sur les passions de l'amour* saw it, passion is a capacity that awakens in us and torments the heart even before it has found an object, which explains why it fastens on an object so quickly and why it can so quickly fade. For while we have seen passion confine itself to only one person who both symbolizes and nourishes it, as happens with love, we have also seen it change objects without losing any of its force. This was the case with Pascal. His capacity for ambition was transformed into a capacity for love, and in the last years of his life he exhibited the same zeal in seeking mortification that he had once shown in striving after excellence.

Now, when carnal urgency, which gives rise to the passions, is mingled with a dimension conferred by the mind (which makes passions nothing but mind, Pascal says), what relationship can we envisage between them? Would passion produce anxiety, the body being too frail a repository to sustain the weight of the mind? Would this invasion not strain, even breach it? Or is it the inner drive, this vital excess, which brands its message of empty restlessness on our very organs and consciousness, and thereby generates passion? It is impossible, I think, to choose between these two contradictory explanations, and this is precisely what makes for ambiguity of feeling. Every affective impulse encompasses both the spiritual and the physical, eternal desire and carnal shudder, the very high and the very low; but we are unable to find where they are joined, for the seam is where they mingle within feeling itself.

The ultimate objective of passion, whether it be ambition or love, or whether it be both of these at once —for love always aims to conquer and to enjoy exclusive proprietorship, as ambition always hopes to unite lovingly with the object of its search—this objective is immediate possession, although the wish is not apparent or even easily detected. This twist accounts for the paradox of passion; the divergence between the object of passion and the feelings it inspires would drive the impassioned man to despair if his very condition did not blind him to it. If he could perceive the divergence, his response would be in the nature of suffering, but it would no longer be passion.

As I suggested earlier, we might ask whether passion must absolutely have an object. Is it possible that we need not wait to experience this or that driving emotion but are born with a "passionate nature?" Does anxiety become so unbearable that we find a release by concentrating it on one fixed point? This, perhaps, is why we must hate this man, love that woman, yearn for this advantage, fear that danger. Once we have conceived a definite object to which desire can attach itself, passion seems to be an effect produced by that object; the object polarizes our turmoil and could, we believe, assuage it if ever our desire could be sated. However this may be, passion can never be completely explained —not in terms of its goal or of its content or by any explanation the subject himself may offer. The psychologist merely describes its mechanism; the novelist merely retraces its history; the moralist merely points to its defects. Each of these observers is aware that if the passionate man were to be granted the money or power or possession he desires, he would forthwith discover some new lack in what he has. So the object of passion is only a sign, a symbol. I would say, further, that while the object does provide passion with its opportunity, the body and our inner turmoil supply its fuel. Form and content come from some other source, however, and I am inclined to look for them in man's need for an eternity concentrated in the instant. By virtue of this shift, passion is spiritual in essence and is very close if not to reason, as Pascal claims, then to intelligence, which is a more precipitate, avid reason that has as one of its most striking characteristics the tendency to try in all things to curtail the period of waiting.

Let us also observe that the passion to possess instantly can very quickly reverse itself and be translated into a hunger to destroy. If hate follows so easily upon adoration, it is surely that hate and love are obverse aspects of the same fundamental emotion. Why do we want to destroy the thing we adore (or want at least to have it at our mercy) if not because we wished to ravish it in one timeless moment and, having failed, take revenge both on ourselves and on it? Strange as it is, by the act of annihilation, or whatever other name we wish to use, we cause emptiness to follow upon full-

ness and death to succeed life; such an action is the inverse of creation, which it thereby resembles; in any
event, it allows human beings to fill time with an
achievement analagous to that of a demiurge.

Such analyses of urgency—which is associated with
the body—and of passion—in which the spiritual element is dominant—should help us, I believe, to see
more clearly what pleasure is, for it is to be found at
the confluence of these two. Most of us are not sick
enough to succumb to the sufferings the psychiatrists
describe or violent enough to experience consuming
passions. We glimpse the gross image of eternity that is
offered us by the senses or by vanity on a more common-
place, vulgar level.

Now let us see whether the impulse to idolize the
present may be detected in a political or moral guise.
Not often, we find, and the reason is easily guessed. It
is difficult to socialize pleasure, because pleasure conflicts with rational and moral effort, and so it is a threat
to the stability that society needs. Thought is impossible until such time as pleasure is abolished or rejected,
or at the very least moderated or deferred to some future
time. The basis of a country's economic policy is to
prevent the immediate consumption of the national
resources. The problem is to establish enough confidence in currency, credit, and investments that the
public will agree to defer consumption of the fruit of
its savings. The state has as much interest in this as
has the individual, for all governments under which
private property exists can base future taxes only on the
savings of its citizenry. Similarly, moral thought cannot
be born until such time as we agree to forego momen-

tary pleasures. Even if we consent to such a renunciation only with a view to guaranteeing superior pleasure
in the future, the sacrifice still acts as a check on present
pleasure and defers the moment of compensation to
some unspecified future time.

However, let us set aside arguments based on econom-
ics and suppose that, for example, renunciation is made
reasonable and even necessary by the certainty of our
imminent death or, how shall I say, by the certainty
that the whole human race faces total extinction. In
this case, would it not be our duty to enjoy ourselves
unrestrainedly? Would madness not be the measure of
wisdom, since only folly would let us exhaust all the
pleasure that the universe affords in the few hours
remaining to us. When Renan was an old man, he once
said that if we were to find ourselves face to face with
sudden, sure death, only our animal nature would
speak up, and without a second thought the world
would toss off a potent aphrodisiac that would allow
it to die of pleasure. Men would meet death with a
sense of exalted adoration, with a sense of offering up
the most perfect prayer.

In Romantic literature we find the idea of a para-
disiacal coincidence between the most extravagant
heights of pleasure and the loftiest moments of spiritual
life. However, precisely such notions as these make it
evident that absolute pleasure is related to absolute
death, the one leading to the other, which is, in this
sense, its fruit. These ideas would be valid only if we
were able to pass without delay from the total presence
of sensation into a state of eternal nonbeing. Such a
prospect can linger on the horizon of thought as a

hypothesis tempting to our egotistical selves at moments when the probability of a Beyond also seems to have been shattered. The bond between pleasure and death can be evoked by those who want to give immoderate flavor to the passing moment. The idea of the brevity of time serves as a seasoning of pleasure, but enjoyment would turn into suffering were it to last, and would perhaps not be bearable if we knew we could have it without hindrance or risk.

Imagination can construct dramas from such ideas that will always enjoy great success with the majority of people, whose metaphysical sense is most effectively aroused by the language of voluptuousness and death— the double screen that conceals the eternal infinite. But no doctrine can be derived from these experiments that one could submit to men's reason, unless it be the maxim, "Let us eat and drink, for tomorrow we shall die." Carnal man alone is speaking here. It is impossible to conceive of human society or human life existing with this outlook or under this law.

On Dissociation

It takes no thought or effort to contaminate the temporal with the eternal. It is enough simply to be alive. Like so many parasites being continually reborn, inner reflections of an eternalized future or an immobilized past or a pleasure-filled present appear and reappear in our consciousness.

But if life amalgamates and contaminates, there is one faculty that dissociates; for want of a better name, I will call it mind. And for purposes of the question that concerns us here, to dissociate may be understood as separating the temporal element from the eternal and never allowing them to become intermingled.

Simply by dint of flowing along its way, life infects the eternal with the temporal and composes a bastard mix-

ture of the two. In order to separate what is thus con-
founded, we shall turn to the entirely opposite function
of the mind. As Plato showed us, the mind distinguishes
among essences so as to preserve each in its purity. And
so, at this point, we will not mingle, we will not contaminate; we will distinguish and we will separate. We
will isolate the eternal element that resides in current
life, and we will let the temporal go. Actually, this is
the solution that the mind pursues by preference when
aroused to its highest degree of subtlety. Ever since
Christianity spread through the Western world, philosophy has tended—at least when pushed to term—to
dissociate the eternal from the temporal and to reject
the latter.

The most striking contribution of the cults that
spread throughout the Graeco-Roman and Near Eastern
world at the end of the Greek era was to advance the
idea of the soul's salvation and, by extension, to call
men's attention to the fact of death, a fact which philosophy habitually disregarded. Many things conspired
to bring about this new orientation: man's invincible
sense that there is a beyond; the idea of remunerative
justice; the kinship that popular beliefs ascribed to gods
and men; a moral anxiety that was heightened as the
human conscience became more sensitive. Philosophy
itself contributed, for it had succeeded in freeing man
from the bonds of both cosmos and city-state, in the
architecture of which he had once seemed to be an
essential keystone; henceforth, he was without base or
bonds—like Plato's Love, more demon than divinity.
But let us think for a moment about the idea of sal-

vation. It was to give hitherto unknown meaning to two pairs of concepts that philosophical thought had generally set apart, as if they related to different, noncommunicating areas. These were good and evil, on the one hand, and, on the other, time and eternity. Henceforth, it would prove impossible to separate them. Good and evil would no longer be defined by a single canon, whether social utility or balance or beauty. They would be considered as two paths that lead to two eternities of contrary kind—two paths ending in two modes of life.

Time becomes, then, a place in which something occurs, the place of an action that engages eternity. But this is a very troublesome concept. Must we still think of man's changing from a temporal to an eternal being in terms of some sort of voyage? Would not that order of eternity still be some kind of future time? Might salvation consist, perhaps, in a very different orientation, the secret of which it would be philosophy's task to explain? From the outset, philosophy had had a practical side that was concerned with good conduct. But it proposed rules of conduct that were derived from speculation about the cosmos and the order of the beings that compose it. The wise man established a hierarchy of essences, even making the bonds that unite them to each other completely intelligible, and thereby found his own place, balance, and peace.

There is no place in such a view for absolute good and evil. Need there be, after all? As we know, the Socratics did not easily accept the idea of sin; they tended to diminish evil to the status of a mistake in direction, proportion, or dosage. In these circumstances, man's good conduct did not consist in struggling with the self at all. The basic distinction between good and evil was overcome by placing oneself so that the mind directly confronts the good and therefore is unable *not* to submit to its own salvation; knowledge dissipates temptation, so that the mind is free only to win salvation.

Philosophic tradition, then, implicitly contained a kind of wisdom that could make the religious life useless. It allowed men to dispense with religion or, at least, to transform it. The mind never functions with fuller power or competence than when it is translating fundamental ideas of religion into the language of pure thought. In our Western civilization, doctrines of dissociation appear at intervals alongside the path that Christianity has traveled.

It is fairly easy to determine when dissociative doctrines will find the most favorable terrain on which to develop: an element of protest lies at the root of all thought and provides the faint stirrings of revolt that are needed to awaken it. We can readily see what in religion will always jar the philosopher. The salvation of the soul is bound to the observance of various social practices that would seem to be explicable, contingent rites; reason is made subordinate to the obscure yet seemingly prescient knowledge of a priesthood; most particularly, what should be seized and enjoyed in the present must be deferred to the future; the survival of social man after death must be assured; virtue is corrupted by extending the hope of reward; the animal instinct is exploited while at the same time an attempt is made to eradicate it; man's zest in temporal pleasure

is replaced with the hope for eternal joy; his impe-
rious dreams are deferred to another time and place and
therefore do not become the mainspring of radical
reform but are transposed and sublimated—these are
the main charges of the prosecution in the case that has
been argued against religion through the ages. It is difficult to deny that even the purest religion may lead
to and at times actually does exhibit deviations such as
these; this is most likely to happen when religion is worn
as a cloak but has not really found a place in men's
hearts.

Accordingly, we may anticipate that when a thinker
who eagerly aspires to scrupulosity and autonomy
(namely, that reason be sovereign in the sphere of mor-
als) comes in contact with more or less debased or
corrupt religious circles, he will be drawn toward a
philosophy of salvation that contracts no alliance with
hope. He will dissociate the eternal from the temporal
so that that eternity which is of the present may be
enjoyed in time.

We first detect a kind of dissociative thought in Alex-
andria, in the third century A.D. At that time ancient
civilization lacked a strong spiritual principle. The religion of the city-state had become a religion in name
only; even the religion of Rome did not possess enough
vitality to spread through the Empire where, in any
event, the center of gravity had by then shifted to the
East. The East was producing a profusion of ambiguous
cults that combined religious feeling and frenzy. Juxta-
positions and, what is worse, amalgams of belief prevailed everywhere, yet in a world that did possess a

center of political unity no spiritual principle proved potent enough to create religious unity. Neither the mysteries of Isis nor the cult of Cybele and Attis nor even the cult of Mithras succeeded in doing so, any more than did the various philosophic systems.

The problem was to reconcile the demands of the populace with the aspirations of an elite, to enlarge a pagan mystery until it encompassed the universe, and to find a symbol that would be all-expressive. (The Severians believed they had resolved the problem by making the sun the universal god.) But there has been no period in history when contamination prevailed so powerfully or when equivocation came so near to being venerated.

We might assume that such a period would favor the spread of Christianity. Until then it had survived largely in secret, but potentially Christian doctrine could answer the confused hopes of the multitude without offending thinking men. Favorable conditions were accompanied, however, by a host of refractory influences. Also, the early Christians faced a dilemma that will always confront proselytizers: Must they renounce the world in order to preserve their faith? Or must they go out among men to convert them at the risk of being corrupted themselves?

Christianity was not daunted; it moved out into the world, and the life and stability of the new religion were linked to a tireless propaganda. However, especially in that syncretic period, it was inevitable that around official, orthodox Christianity sects should spring up (the lines of separation were not always easy to trace) in which elements growing from the Christian

germ failed to assimilate alien matter and were instead
debased by it. Foreseeably, these bastard amalgamations
of pagan images and Judaeo-Christian concepts prevailed the more powerfully over men's minds the more
fully they could satisfy conflicting needs without requir-
ing of humanity the sacrifices that it so begrudges.
Perhaps I have just offered a definition of Gnosticism.

Outlines are more sharply defined when a shadow is
cast over an uneven surface; similarly, Gnostic doctrines
that resulted from the projection of Christian dogmas
on alien ways of thought highlighted several characteristics of Christianity itself. As they were removed from
their original environment, fractured, and distorted,
typical features of Christianity appeared in a grotesque,
intellectually discreditable form. The sustained execu-
tion of a proselytizing plan broke down, and the doctrine spread by fits and starts, with catastrophic results.
Opposing principles were united without regard for the
differences between them so that the one, in effect,
imprisoned the other; unity came to be confused with
fusion, and distinctions with contradictions. God the
Creator was set apart from the Supreme God; souls
were locked within the jail of matter. Contention,
quibbling, and deterioration were the rule.

These were the conditions that prevailed among the
Gnostic sects that Plotinus, to his dismay, encountered
on his travels. It has happened often throughout history
that philosophers have known the Christian religion in
one or another aberrant form, which is not surprising.
If the essence of "wisdom" is to propose a doctrine of
salvation in which the mediating term is supplied by
reason alone, that doctrine can be nourished only by

what seems most contrary to it. That is to say, there is a harmony between the aspiration after perfection and an encounter with corruption. The insistence on complete purity finds sustenance in a critique of error. It is both strengthened and justified thereby, and slips imperceptibly from indignation into affirmation.

We find Plotinus challenging the notion of a material eternity, an eternity conceived as an entrance into some new heaven and new earth. He is apprehensive when he sees divine action made subject to history; he is scandalized to find it represented as the work of an artisan who thinks in terms of specific objectives which he executes, one by one; and he is appalled to discover the demiurge invested with human emotions, such as the desire for glory.

Simultaneously, he criticizes everything that serves as the structure for ethical salvation, namely, evil, time, memory, individuality, and even perception. Does evil exist? How could evil be introduced into a world that is the expression of first principles? Is not what we call evil the aspect that a universe necessarily assumes when its inner law has forced it to deploy itself in a space-time where its parts, having become discontinuous and even discrete, can hold together only by mutual opposition?

Does freedom exist? Yes, beyond any doubt, since we must speak of a tiny "upset of equilibrium" and even of "audacity," perhaps, in order to explain if not the procession then the incarnation of souls. But in the specific case of the fall, does the soul not surrender to a vertigo which is outside itself and which is a law of nature?

Does memory exist? Above all, does it survive after death? This is unthinkable, for we would have to count as many memories as there are faculties and especially genres of life. Memory, therefore, is bound to historical existence: as we divest ourselves of our social personalities, or of our cosmic wrappings, we are less and less conscious but more and more ourselves, and at the very end, when the memory of past lives has been eliminated, we are pure being. In these circumstances, personality being bound up with memory, has no absolute existence; what does exist is the various roles we play in time, under the influence of astral conjunctions and our duties.

As for perception, it does not consist so much in apprehending something outside ourselves as in allowing an indivisible thought to unfold in an inner language and be reflected in us as in a mirror, at which point consciousness can grasp it.

Does time exist? Even less so, for it is bound to the one essential evil—the inner void, the malaise, the pain—that is introduced into the world soul and spurs it on to search for that which it already has. In all events, souls themselves do not exist in time; only their sensations exist, their passive states, and their sufferings. Therefore, the only thing to do is to escape, or rather finally to become aware of another order of existence, which is the true one.

Several times Plotinus experienced states in which the soul seemed to reach its homeland. What it experiences at such a moment is not thought but contact, a palpation that we have no word adequately to describe. But I would ask whether such rare states are not related to what his teachers, and Aristotle in particular, seemed to indicate were the final limit to which dialectic must ascend once it has purified being of all accretions acquired from without during its passage from the supreme and empty genus to the most differentiated species?

Logical mysticism had already appeared in Aristotle. By progressive abstraction, thought reduces the many to the one and also abolishes time, which is composite. By discovering and establishing the universal, thought delivers us from time and place. By a kind of inversion, thought seizes from behind the mobile order of causes the immobile order of effects, which the former unrolls as if backward in time. But since necessity is one with eternity, since intelligence is one with the intelligible, since the thought of the eternity of the object merges with the eternity of the subject of the thought, man can be eternalized by the act of thinking; in this way, time is abolished. A close bond is established here between logic and the doctrine of salvation. Plotinus undoes this bond.

His guiding idea is that one must search for eternal life not beyond time but in time itself, following a path that is at once dialectical and mystical: dialectical in the sense that the chief task is to find via thought the unity, the immobility, the eternity that is in things; mystical because the soul must make this journey by passing through all the stages of spiritual life to arrive at being one with God. Perhaps Plotinus' most profound idea is that this is not a dual but.a single effort; that is to say, by striving to transform the soul's successive states into moments of thought we will truly win salvation.

To support this ethic, Plotinus wove a metaphysical background that we may well call eternal, for it is always implied in doctrines of eternal salvation. He assumes that being comprises all that is required to condense time and reduce it to eternity in the present. Creation is inadmissible, however, for creation would place an imaginary abyss between the first and second beings, the creating and the created, and would split being into two opposing forces. What is more, it makes our end as well as our beginning depend on the will of another; it ascribes caprice to God, who is necessity itself. Worse still, it assumes consciousness in God, since we are then forced to envisage God in the image of man. If the world has issued from The One, this must be understood as a god's issuing from a god—be understood, that is, as generation. If such generation fails to explain the diffusion of being, the aspiration of the soul, and the resistance of the will, let us adduce a fall, or rather, some mysterious flowing out that is owed to both filiation and flaw.

It may be objected that such a concept of procession makes no sense because it adds to pure being all the flaws, deficiencies, and lacks that I have just enumerated. Indeed, we have here a capital mystery, but one which we may solve by conversion: we travel back over the same road, we redeem the fall. Or better, without moving—for in an eternal world motion is impossible— we allow the clouds to dissipate; having interposed themselves between us and ourselves, they could mislead us into assuming that we were beyond or outside

57

Being, in a real and demonstrable time. But we have not ceased being in God and of God. Nothing at all has happened.

There are, then, two beings in each of us. One beyond time, one in time. We can understand Plotinus' irritation when people talked to him about *nous,* and his reply: *"Nous?* Which? The one that is there, or the one that suffers and becomes in time?" We are, he remarked, amphibians.

Spinoza lived in an intellectual and moral milieu that favored dissociation, a world of such unwholesome alloys that could only stimulate the urge to purify the bastardized forms of thought which then prevailed.

In the Low Countries, a struggle between contending, albeit related, religions—Catholic, Protestant, and Jewish—forced each to define itself clearly in relation to the others. The Jewish community of Amsterdam displayed the abiding characteristics of Judaic organization during the Diaspora. Joined to the world by commerce, the Jews were separated from it by their race, their sacred literature, and their group aspirations. They had always tended to hold themselves aloof from the societies among which they lived and to shield themselves from contacts that could corrupt the Judaic spirit. They had sought to preserve that spirit by practicing the Law to the letter; yet if the Law had to be kept faithfully, in order for them to survive it had also to be betrayed, which led to the casuistry one finds in the Talmud.

Such tendencies, natural enough in the Diaspora, were heightened and exacerbated in the Jewish community of the seventeenth century, when the religion of Moses and Ezra was besieged by both Christians and atheists, and threatened to collapse before so many enemies. It had no way to maintain itself except through literal adherence to the Law—unless it were to generate a pure, vigorous, and uncompromising spirit, freed forever of orthodoxy—what, in other words, Spinoza was to be.

Spinoza's was a clear, incisive mind, disciplined and discriminating, and seemingly made only more lucid by illness. To such an intelligence, the prevailing duality of spirit and letter, leading inevitably to equivocation, was shameful beyond endurance. The scriptural interpretations offered by the rabbis were, in his view, nullified by the Bible itself. Extend this point of view to all religion, and you have the protest which nourished Spinoza's thought. To him religious practice seemed a futile, sterile surcharge; he found cult a tyranny, or as the only book he published put it so succinctly, a "theological-political" system. As he saw it, religion faced the challenge of honoring its pledge of rigorous purity. Simplify, eliminate nonessentials, replace external forms with inner significance—Christianity had once done this for Judaism, and this was what Spinoza sought to do for Christianity.

For this purpose, he needed a rigorous discipline that would enable him to strip away fantasied and anthropomorphic accretions. Not surprisingly, he turned to mathematics; he found it the one method that deliberately aimed to exclude all corrupt modes of reasoning which reverse the normal order of thought by deriving premises out of wishfully established conclusions.

Any deduction based on the principle of human purpose had to be completely eliminated. Mathematics was most helpful here. Intention has no place in its indifferent, severe emptiness. If faithfully practiced, such a method will erase our tendency to see what we should like to see rather than simply to learn what is. Descartes had also used mathematics but in the form rather than in the substance of his thought. It remained for Spinoza to push through to term, that is, to use mathematics without reservation and even, when necessary, use it as a method of exposition, as he resolved to do in his last work.

Spinoza developed this philosophy to its highest power, and it is not surprising to find here again the assumption that the Greeks, in their wisdom, had made with regard to salvation and that had been eclipsed but not eradicated by ascendent Christianity, that is, that intelligence saves without any mediation other than the inner word, or rather, the mind that has no need of the word.

Spinoza retains an idea which is basic to the philosophic tradition, namely, all that belongs to the domains of feeling, love, and mysticism, or, to use the Hebraic and Pascalian term, to the realm of the "heart"—all this must be ordered, systematized, espoused, and proved by the intelligence, indeed by the most exacting form of the intelligence, which coincides with the necessity of nature. Spinoza has no difficulty in identifying the ascension of the soul into the order of God's love with its elevation into the order of the knowledge of truth. For Plotinus, the identification of the soul with God always remained a problem or at least a mystery; it was almost a matter of faith. For Spinoza, this identification or, rather, this awareness of identification with God can be the subject of an intuitive experience and therefore is capable of providing its own proof. It is enough to raise oneself to the third level of knowledge, that is to say, to a mode of thought in which man is able not only to apprehend necessity via passive resignation, like the Stoic, but where he is also able to understand necessity and to establish himself within it, so to speak, by engendering its law. In this way, the soul is saved without having to rely on imagination or hope and can dispense with the idea of freedom and of purpose both in man and in God.

The Cartesian meaning of the word "imagination" always rejects the notion of a power that subjects us to time and that seems all the more deceptive and dangerous when it places time in an aftertime by forging the bastard concept of immortality. "Life after death" is an illusion. The mind is held in thrall by this image and cannot rise to a true notion of eternity. But we do not pass from our precarious and fluid life to eternal life as a result of some extrinsic accident; we raise ourselves toward eternity by a mutation that is brought about by meditation on necessity, and that establishes us suddenly, as it were, in another life.

There are two ways to conceive of things as being current. We can relate them to a well-established time and place; imagination and even reason visualize them in this way. (Here I mean the ambiguous reason that is still close to the image and espouses the necessity of the universe without being able or wishing to think of it as being in God, where it is eternally.) But the mind may also conceive of things as contained within God Himself, as if deriving from the necessity of His divine nature; seen from this perspective and in this light, the living soul and the present moment envelop eternal and infinite essence; things are real—or, rather, they are true. We feel that we are eternal, but not because eternity or any religious experience conveys such an impression to us. On the plane to which we have now ascended, these words, which denote palpation and duality, no longer have meaning: demonstration has become a matter of observation; it is no longer a question of our feeling or hoping or believing. That which is light has no need to see. *Per aeternitatem, hoc est, infinitum existendi; sive, invita latinitate, essendi fruitionem*: eternity is the possession of being.

In other words, a victory without a battle. No suffering, only salvation. There is no longer any struggle between good and evil; there is no real distinction between this life and the life to follow. True wisdom raises us above the perspective with which these two images are associated. So long as we visualize good and evil as acting on us from without, as proposing alternatives for us freely to choose among, we assimilate them with the passions, or at least we invest them with the same kind of causality; we live under the yoke of hope and fear, and if we are able to cure ourselves in these conditions, it will be by drawing from evil the strength we will use to vanquish it. No doubt this is how we all start out and how most of us continue to function. For most men, good conduct and vice are based on the same principle. What makes us give up sensual pleasure is the profound bleakness inherent in it. The pursuit of wealth and fame does not immediately cause such a reaction of distaste, but because the pursuit is insatiable, eventually we stumble into frustration. Not even our desire for immortality is free of corruption; the expectation of a reward is itself a passion, since it places us outside ourselves.

True salvation consists in lifting ourselves above such images; we cannot suppress them, so we must exercise judgment and dismiss them, together with measure, time, and number, as imaginary. Redemption—a new birth—does not mean that we cause good to triumph over evil by making one kind of love prevail over another, as St. Augustine and his followers would have it; man is redeemed by ridding himself of the idea of love of good, of evil, of self, and even of God, inasmuch as these loves are external to him, in order to find the one unique love—the love with which God loves Himself eternally—and to know himself as one moment in this eternal circuit.

Spinoza never maintained that man could establish himself in eternity straight off, as it were, as the result of a simple decision, else why would he have written a book telling us of our bondage, of how difficult the journey is, and how rarely the goal is reached. Life is lived on two levels. We are like sleepers who are in thrall to their dream images, but who nonetheless realize that they are dreaming; they deny their dream but they are unable to wake up; they come finally to understand the laws of succession in dreams and are able to escape the false euphoria and anxiety that such images arouse in people who are ignorant of the nature of the dream experience—yet they cannot understand why they dream.

Have we, with this, come back to Plotinus? He also believed that salvation was not to be posited beyond time; he did not concede the reality of evil or of freedom; he also thought that the soul is saved by becoming more and more aware of its identity with God. But between Plotinus and Spinoza the world had changed. Plotinus was never able to conceive that in redemption the supreme principle acts in some way on the soul that is turning toward it. Spinoza, on the other hand, wished to preserve the essence of Judaic tradition and perhaps even of Christian thought. The man who called Christ *summus philosophus* could not fail to deal with the idea of grace. Here as elsewhere he applied his method of radical separation between the letter and the spirit so as to preserve only the essence of grace in his concept of it.

And what is the essence of grace if not that our love for God is only the effect of God's love; that the soul is saved not so much through its awareness of the love it bears God as through its awareness of the love God bears Himself through it. We can stop loving ourselves imperfectly only if God loves Himself in us; as a consequence, detachment from the passions is not so much the means whereby we obtain grace as the sign that grace already dwells within us. Certainly the essential theme of Christianity seems to imply that life is separated into two planes—an eternal plane where everything is established forever by the predestining choice of divinity, and a temporal level where everything is subject to revision and compromised by human freedom.

Yet not even the most exacting Christian thinker—
Jansen, for example—who has been eager to distinguish
between these planes has succeeded. Two paradoxical
virtues, faith and hope, insure the indissolubility of
the human twin—eternal man who has his being in
God's thought, and temporal man who hesitates, makes
choices, and falls into error. But then Spinoza arrives
with his dissociative logic, and time is not proved but
rejected in favor of divine necessity. The idea of grace
survives in the sense that God alone, who does not love
men but who includes them in His love, is able to provide men with the momentum that will set them free.

We must not think of Spinozistic salvation as resulting from any human initiative; it comes about through
a kind of eternal prevenient and indispensable illumination. With this concept, Christianity attains to a
fuller and inwardly more rich expression than at any
time since Plotinus. Assimilation replaces analysis, a
lofty sympathy replaces disagreement, yet all effort is
directed still to the same end—to transpose new values
into a universe where time is stripped of being—and
this, I think, is possible only through dissociation.

I have now to describe a third position that to me
seems characteristic of our times.

Let us apply the method already used and ask first
how spiritual corruption, which seems to spark sin and
at the same time to be the source of regeneration, will
now manifest itself. In our day, has the cause of the
Spirit ever seemed to be compromised by the very people who represented it? Indeed yes; we need only look
at the bourgeoisie, which has been the dominant class

in Western society. This middle class has had a kind of spirituality, a kind of moral philosophy. It has allowed for a distinction between good and evil, conceived now in biblical—or more properly speaking, in religious—forms, and again in separate, lay terms; above all, bourgeois ethics have assumed that the good is synonymous with goods, honesty with utility, and following the dictum of Montesquieu, that religion assures us both a heavenly hereafter and enjoyment of the fruits of this earth.

It is no part of my purpose here to show what sound elements this doctrine might contain. It is close to the earlier Judaic idea of temporal retribution. The lure of wealth becomes less noxious when the acquisition of wealth is made conditional upon virtue. Anglo-Saxon societies have been built upon the maxim that the rich man is beloved of God who rewards him for his labors with a fortune honestly earned. Calvin's famous question about the signs of divine election was answered in the same sense: What more constant, palpable sign could there be than material well-being and security? Franklin's morality leads directly to Bentham's. Every honest middle-class man is a Franklin in his heroic moments and a Bentham on weekdays—a Franklin in the bosom of his family, before his own children, but privately or at the office a Bentham.

The corruption of morality leads to pseudo morality. To the extent that society is based on the pledge, on the promise, on virtue, or on a religious creed, it attributes to honesty—let us say even to the semblance of honesty—a utilitarian value. Superficial honesty happens to be one of the most powerful springs of social advancement. A man need present only an honest exterior to reap the profits of public esteem and respect, not to mention the advantages of confidence and credit. And since shrewd, unprincipled action retains its immediate advantage, particularly in a society healthy enough for wrongdoing to remain the exception, we can guess what the supreme temptation will be. Why not preserve just the semblance of honesty and behind this screen pursue a policy based on competitiveness and utilitarianism? More clever than Callicles, who disgraced himself in the eyes of the philosophers, man would stand to gain on both scores.

The deviation here is that honesty is reduced to a matter of appearances, and earthly goals are pursued behind a screen of spirituality; seeming and being are dissociated—in a word, men lie. We must admit that middle-class societies have been prone to this kind of duplicity. How often have they preached one standard of conduct and practiced another; sexual, business, international and colonial political morality have all suffered from the disparity between appearance and reality.

When falsehood becomes so deeply rooted in us that we are unconscious of it—when men speak the language of eternity but live in pursuit of temporal things —then we must expect that dissimulation will become the law of their inner being; if men are led to translate their instinctual drives into the language of virtue and metamorphose them without correcting them, they succeed only in transposing or, as Freud put it, in sublimating them. It is scarcely surprising that such a society proposes to remedy matters by forgetting about appearances: let a man be what he is, and let all rules of conduct go by the boards. It is convention that creates the offense; if trespasses flourish, the law is responsible.

We can distinguish a second and parallel deviation that affects religion rather than morality. Nietzsche characterized it very well as *ressentiment* (resentment). When we try to define how it impinges on time, this is what we find. The resentful man is quick to assume that people who use time for their own pleasure are lost souls; however, because he is more capable of fear than love, we may suspect that even as he condemns, say, the profit-seeker, he also envies him. If you will listen to him, he will tell you that time has no value in itself; it has value only as preparation; time is granted to us so that we may deny ourselves the advantages it offers and so that we may multiply, or at least savor, our tribulations in such manner that human life has no purpose other than to procure us rights in the hereafter. If this were true, then the organization of society would foster an attitude of resignation. I would call this attitude a morose pseudo eternalization.

Hypocritical moralism and this surly pseudo eternalization can appear together in certain milieux—for example, in circles that are attached equally to property and to piety, but that have lost all sense of the spirit. Such a society existed in the nineteenth century. Its corruption gave rise to a protest, as has happened before, that took a contrary direction, yet the protest was shaped by the abuse it undertook to correct. The dissociation of appearance and reality was to be eliminated by canonizing the natural man; at the same time, the dissociation of time and eternity was to be avoided by seeking eternity in time, although it was to be sought in this instance in a new, more complex, and more immediate way.

Several of the great dissidents of the period thought along these lines. As Rousseau their prophet had done before them, they combined consistently vehement and occasionally cynical protest with a profound sense of humanity. We hear this protest in Nietzsche, Wilde, and in Dostoievski, who was perhaps the most pure and humane of them all, a man who knew how to endure suffering without pride. I will try to characterize them all in terms most accessible to me as a Frenchman, that is, through the work of André Gide.

Everything predisposed Gide to be acutely aware of the grim flaw inherent in pseudo eternalizing: in his strict upbringing virtues were negatively expressed as prohibitions, duty often consisted in perpetuating social position, and the natural impulse was likely to be equated with sin. Some measure of dissimulation was necessary, too, because the safety valve of admitting one's misdeeds was denied, yet the standards to be maintained were sternly evangelical and the goals assigned to life were difficult if not unattainable; the Kingdom of Heaven was to be entered only by "the narrow gate." On the other hand, Gide inherited a nature so generously endowed that he was naturally indifferent to merit or praise. His was a prodigal nature, prone to squander its riches; he showed little propensity for integration and delighted rather in the balancing of opposites; he was more sensitive than most to the counterfeit coin of common virtue.

If we disregard the metaphors, the parables, the farcical and indecisive aspects of this reflective man who was so attached to the inconsequential, and look instead for the general lines of his larger design, we will detect the thread of a theory of eternity in the present which is extremely coherent, less because the author wanted it so than because of its inescapable logic. It is an idea common in the Epicurean tradition: since nothing can supersede the moment of pleasure, man's proper work is to give that moment a new dimension in any of several ways—by the elimination of fear, or by a sapient reminder that death is imminent, or by the extension of art.

One might ask whether all art, whatever its form, is not Epicurean in essence. Art shatters conventions and by adroit devices seeks to assure us of a connection with the reality of things. Now, why could we not use the senses in a way analagous to art's use of its various tools? Could we not make use of the body to discover the reality that is concealed by custom and convention? Is morality not simply one of the customs that prevent contact between our own being and life?

Such ideas bring us quite close to the position taken by the immoralist. We would have to add only his constant recourse to evangelical expression and to Christian ecstasy and mysticism, which Gide made serve a sensual idea they would previously have seemed to exclude. Salvation was to be sought in time itself, and, as a consequence that we now recognize is necessary, in certain particular moments that possess eternal value. We are introduced to the cult of the moment, in which satisfaction is achieved without intermediary and without delay. We are in the garden of Epicurus, but our

life there is very different from his, since ours seeks to
assimilate the whole mystique and ecstasy of nineteen
centuries, all of them equally inimical to pleasure.

We have been supplied here with a technique for
detachment and for total suspension—detachment, need-
less to say, in regard to morality. Virtue is repudiated
out of love of virtue; in order to escape the distinction
between being and appearing, which engenders insin-
cerity, we must repair beyond the permissible and forbidden, these terms serving to define good and evil.
There are various conceivable ways to divest the idea of
sin of its meaning, and so far we have dealt with two:
one considers sin as ugliness, as an excrescence; the
other severs the possibility of error at the root by dem-
onstrating to the intelligence that freedom is impossible. Gide, on the other hand, comes nearer to an idea
we find in St. Paul's polemic against the Judaists, when
he struck out against the law: sin is caused by the pro-
hibition of sin. However, Paul did not believe that if
law is abolished sin will thereupon disappear, for, as
we see, for him the duel between *spiritus* and *caro*
continues. Gide contends that if convention is abolished, abolished without hesitation or remorse, this will
enable us to recover our innocence. Sensuality consists
simply in considering the present object and the preced-
ing moment as an end and not only as a means.

This is a prime method whereby time may be redeemed; we reverse what time has unfortunately contributed, and we return to an earlier condition not by
a reversal of will but by repudiating convention. We
renounce even reason, which is external to us at least
in terms of the motives it offers us, in order to obey a

stronger reason or causality that is characterized by its opposition to the rational.

We are summoned to act without prior deliberation or sanction; action is swift, abrupt, uncalculated, and it is the more meaningful for not having been weighed. However, we would be mistaken to suppose that sincerity is easily achieved; sincerity demands sharp insight and an absolute rejection of self-complacency. This view of grace is termed "gratuitous," and through it the first stage of detachment is attained.

The second will tear us away from the society of other human beings and from attachment to individuals. To love a person means that we are still dependent, whereas we are delivered of dependency if instead we love friendship or affection or love itself. And to achieve still greater detachment, to come finally to the point of eluding all affective liaisons, we must become nomads, we must avoid all connection with time and place and avoid all opportunities for encounters or attachments. This will be possible only through a perpetual displacement or at least through a readiness to accept any future whatever; this attitude is termed "availability."

Now the spirit is no more than a pulsing, impersonal will. If it is accused of being disordered, it can always retort that such disorder is less dangerous than an arbitrary order that is not invented afresh each moment; in any event, it is a provisional disorder that prepares for our rewon innocence, But for detachment to be total and for it to become humble destitution, it must be practiced at the very core of sensual joy; it must constantly dissociate joy from itself to prevent desire from

becoming attached to an object, which would dim joy.
Therefore we renounce memory; we abjure everything
that might stimulate recollection or regret, and by so
doing, we invest the moment with duration. Likewise,
we reject every vision of the future, for that also would
vitiate the present. And if within the boundaries of the
present moment we could contrive to have appetite pre-
cede pleasure and tarry in possession of it forever, then
we would know the infinite presence of the present, and
feel the pulse of being. This third degree, which we
might call possession of the third kind, is termed "fervor"; it is almost impossible of achievement, and sub-
sides into melancholy.

On the sensory and affective levels, fervor is the
equivalent of beatitude. By heightening sensation and
inhibiting thought, we have accomplished the cycle
without introducing any delay. We have not deferred
the moment of happiness. *"Et nunc est,"* as Jesus
preached. Eternity is of this moment if only the soul
will abjure itself. And there are no prescriptions. What
peace this is! Belief in the survival of the soul feeds
on this same need of eternity, of course, but the believer
is overwhelmed by despair because he cannot satisfy
this need during his lifetime. "Shall I tell you what
prevents me from believing in eternal life? It is the
almost perfect satisfaction I derive from striving after
and immediately achieving happiness and harmony."

If we strive for supreme knowledge through sensation, will we not again end by separating the instant
into two parts? One part would soar above time into
the continuity of pure aspiration, and the other would
be debased below time in the discontinuity of sensory

fluctuations. One part would touch eternity through the ecstasy of deprivation and purity; to put it differently, what would rise from the depths of unsatisfied desire would not be a choking bitterness but rather a liberating sense of ease and refreshment. Yet while one part of the instant would be lifted above pleasure, as it were, the other would be cast back into the vital and temporal; it would be savorful, yes, but faltering, now lost, now found, now lost again. Neither angel nor animal but angel *and* animal—better still, angel because animal, freed of the bonds of time precisely because subject to time, in a state that we cannot call an absolute dissociation of the spiritual and sensual or, on the contrary, a perpetual oscillation between the one and the other—this is the new man Gide proposes.

If the instant is divided, would not division also manifest itself in the problem of love? The Judaic idea of the complete spiritual and physical unity of two beings in a single flesh has no more curious opponent than Gide. According to him, love cannot *not* be divided into two parts, and this is where one sees, as in a magnifying mirror, the dissociation love imposes on duration. To his way of thinking, only two kinds of love can exist: first, love that is unrealizable and that can be fulfilled only through renunciation; second, the vulgar, mercenary, and immediate love, which acts upon an ordinary passive body without regard to its sex. These two kinds of love can alternate: one can redeem the other; however, they can never be conjoined or make use of each other. Pleasure is the more pure, love the more perfect, if the heart and body remain disengaged.

By a strange paradox, if these two loves are not

focused on the same object, the lower can sustain the higher; as self-respect diminishes, the veneration of the beloved object is augmented. The man portrayed here is *animus* on the one hand, *sensus* on the other; he is not *cor*. Through aspiration the soul lifts life above time, while a spasmodic and shuddering sensuality fragments time into a series of instants, thus making its existence analogous to matter. The life of the heart, on the other hand, owes its profound and rhythmic continuity to restorative memory and to the committing promise.

The three positions we have just examined differ so greatly that one is tempted to reject the idea of there being any analogy among them. Yet this pronounced degree of difference will allow us to single out an element common to all. In the case of a strong resemblance between things, we are most likely to learn about each by assessing what it is that differentiates them; contrariwise, when their differences are striking, some hidden resemblance is most likely to prove instructive. By evaluating that which is contrary to appearances we can be led to a deeper truth.

The first resemblance among these three positions is that each denies the widely held concept of moral option. This concept presupposes that man can choose, horizontally, as it were, between two paths that lie open before him on the same level. The contrary view is that this same choice is made vertically, between two planes, the more visible of which is only apparent while the other alone is real.

Let us designate the effort required in each instance

by different names. Let us say that the first is a moral effort; it is an exercise of will that continues throughout life. Habit can lighten the effort; habit can even enable us to make the effort, if not unconsciously, then at least without difficulty and without having always to plan it afresh. The import of the effort is not diminished thereby; it is perhaps all the more pure and praiseworthy by the degree to which it tends to become second nature. The best man is the man who no longer needs to struggle so hard to be good.

I call this a moral effort because it is counseled and often even commanded by what, since it regulates custom, we call moral discipline. However, as Kant perceived, there is in each of us an effort of will—it is perhaps the guiding spirit of the will—which we denote "intention." Our whole being is engaged in this effort. Yet even when intention is enlightened by intelligence, its nature remains different and distinct from the whole being, for it involves a choice, and in making a choice we reject or eliminate or would like to annihilate that part of ourselves which opposes the choice. This recalcitrant part remains operative, nonetheless; it is never reduced to mere appearance. This negation of the self by the self, a negation that rejects but does not abolish, is the essence of sacrifice.

In the second instance, we have to do with what I shall call an ethical rather than moral effort. It was Spinoza's expression, you will remember, and I shall use it to indicate that what I am about to define is not ordinary morality. It belongs much more in the realm of knowledge than of freedom. Obviously, in this second context, moral effort is not abolished. It is needed

as a preparatory divestment; it is indispensable in ridding us of the inner obstacle that weakens our communion with our being. A veil has spun itself over our
eyes and must be brushed away; our spiritual percep-
tions suffer from a kind of sickness, a languor giving
rise to an illusion that must be cured. Once this has
been done, however, we shall not call upon the will to
work our salvation. Salvation will come to us from
that part of ourselves which is capable of knowing
pleasure; whether this is intellectual intuition or sensory or nonsensory palpation really does not matter.
Whether contact is made via the intelligence, the senses,
or the spirit, the problem is not to attain a goal in some
nebulous future or to make ourselves worthy of it by
self-sacrifice. Instead, we are asked to take cognizance
of what is already given and to savor it by possessing
it. Once again, this is not to say that such possession is
easily won or that we ever achieve a lasting awareness
of pleasure and salvation. To ascend to the possession
of the truth of being demands a painful kind of asceticism, for if it is hard to deny sensual temptation, to
reverse our natural vision and perspective is perhaps
even harder. However great this effort, it is not to be
confused with moral effort; in a certain sense, it is even
the opposite, because it is a modification of our vision
rather than of will.

Let us note also that all three men we have considered reject the duality of human nature. For Plotinus
sin is only a blemish on our otherwise virtuous soul.
For Spinoza sin has lost any reason for being, having
been dissolved in his concept of universal necessity.
According to Gide, who goes much further indeed, we

become pure by rejecting conventional morality. None of these men construes good and evil as two values situated on the same plane but distinguishes instead between two levels of being and of life; once this distinction is drawn, duality of the inner man is relegated to the realm of appearances.

Shall we object that duality has not really been challenged but merely transposed? That by replacing two paths of conduct with two planes of truth, one has merely substituted reality for its popularly accepted symbol? The moral and religious man believes that he has an option, the full import of which is that he may choose between good and evil and between their everlasting consequences. The ethical man knows that this choice is only a symbol of the fundamental option by which he decides between temporal appearance and eternal truth.

Even if we were to assume duality of being in both cases, we would have to concede that duality would not mean the same thing in both. In the ethical view, duality would be noetic, not felt or existential. In other words, not the duality of a man divided against himself but of a man divided in two who lives in two noncommunicating worlds. Time in the life of the spirit is dissociated. There is a superior time to which one ascends by conversion and which is eternity. Below this flows an inferior time, which is the milieu of everyday action but which the mind rejects as fictitious.

It is no cause for surprise that the moral and the ethical man lead different lives. The moral man is cognizant of his divided impulses. The division is a source of suffering and torment to him; it poses the problem of how to unite his divergent selves and no matter how lofty his solution, it is always imperfect. Read St. Paul, St. Augustine, Pascal, or Luther, and you will hear the lamentations of the struggling spirit. Duality prevails at the outset of the voyage, unity at its almost inaccessible close. It is just the opposite according to the other perspective. Unity is conferred from the outset. The separation into two parts is an illumination resulting from a philosophical conversion that provides the solution and sets the seeker free. This is not to say that division eliminates all uncertainty and anguish. But if it does contain an element of confusion and anxiety, it is not the pacific torment of a nature striving for unity; it is the perplexity of a mind that cannot avoid apprehending itself simultaneously on the level of time and of eternity.

Strange as the dissociation thus imposed may seem, the concept provides not only aesthetic pleasure but also an immediate advantage. The problems of morality and of salvation are finally consummated in one.

Instead of deferring the moment of happiness, instead of basing it on a sacrifice or a gamble, as is so strikingly true in the case of Pascal, instead of forcing us to risk losing everything and winning nothing, the philosophers we have been studying propose a technique for achieving happiness that allows us to possess what we seek instantly and forever. The distinction between the means and the end is abolished, for the means are the end. The road is the goal, the method is truth.

On the Structure of Time

St. Augustine could as well have bequeathed us several philosophical systems quite different from the one to which he did give his name. Like Plato, he could have associated each with this or that half-real, half-fictitious personality that would have become its symbol, and his thought would have unfurled like a fan instead of being tightly knit and concentrated, or, as he said, solidified.

We can readily detect the mental posture of Plotinus in St. Augustine; the two were so close in their thinking that at one point Augustine recognized himself in Plotinus. Ecstasy had become firmly established as the vehicle of ideal knowledge, and when Augustine wanted to translate his Ostia vision into words, he bor-

rowed the vocabulary of Plotinus. As for the idea that
salvation is won through an awareness of eternal love
loving itself in us (Spinoza's intuition was the same, I
believe), we find this prefigured in Augustine's writ-
ings on grace. There is, of course, the question of
sensual experience, but who does not know the ten-
year-long history of this area of Augustine's life? *Amare
et amari dulce erat magis, si etiam, amantis corpore
fruerer.* More than any other man, Augustine lived on
two levels of being since, according to the Manichean
doctrine he professed, the god of evil could occupy a
part of him without infringing on the realm of the god
of good: *non esse nos qui peccamus.* No one has
described the travail of the senses with greater preci-
sion or wonderment, no convert has ever placed such
explicit insistence on what is normally concealed. St.
Augustine spoke of the sinner with altogether remarkable respect.

Whatever analogies can be found between him and
each of the men considered earlier, they will never dispel a major difference: St. Augustine does not dissociate. And yet seemingly his analysis of time would
have brought him to it. Let me recapitulate a main line
of his thought, which, as it developed, obliged him to
attribute opposing qualities to time.

In the indivisible unity of the present, which St.
Augustine considered a lasting interval (*mora*) and
not a mathematical point (*nunc*), he distinguishes several spiritual directions. One is longitudinal, so to
speak, which he called *extensio*; when this is augmented
to the point of becoming painful for the subject, he
designated it as *distensio*. Following this dimension, the

mind ranges back through the past and forward into the future; it may become distracted and even oblivious by dint of an alternation between stretching and slackening. But there is another, a vertical dimension in which, instead of reaching or wandering, the mind is concentrated and collected. This is the diminsion of attention (*attentio*), and when attention is intense, the dimension of intention (*intentio*). Whatever the terms, they mean, I believe, that our psychological life is a combination of two movements, one of which constitutes what we call time; the other, although it is also in time, shares in a higher reality.

How is it that St. Augustine was not tempted to deduce a principle of asceticism and even mysticism from his analysis of time? By singling out one element, all analysis invites us to isolate that element and deal with it separately. St. Augustine was all the more inclined to such a process of separation because he had learned from experience that the temporal and the eternal are united in a synthesis for which life supplies the means—the solution, that is, of the *peccator*. As the sinner sees it, the current that bears us upward must be converted into sensibility. Man thus becomes a living contradiction: he is a kind of impoverished abundance. In order to compensate for inner emptiness and to prevent what he loves from slipping away from him, he projects the sensory object into infinity, worships it, and thus his error becomes his god. He takes his joy in joy. Things happen for him as they do in the theater, where the spectator enjoys the sensation of suffering without being obliged to suffer. But it is a futile attempt, which in the end only makes him oscillate between moments of self-dissipation and empty plenitude.

What if we could dissociate these two currents? What if it were possible to keep only the current that bears us upward? Then would we perhaps have solved the most urgent problem of human life? I believe that the account of the Ostia vision must be read in this light. Plotinus had taught his disciple not to hypostatize matter or evil, and so had delivered him of the flesh; Augustine was naturally tempted to follow his liberator's example and also to seek the rare moments in time when he might draw near the fundamental principle. He chose the psychological moment, the historical and dialectical moment, and even the place most likely to help him achieve transcendence into time. It was at Ostia, after his conversion, and with Monica, who was near death, at his side. And yet the ecstasy resulted in only a "moment of intelligence," which Augustine interpreted metaphorically rather than as a fragment of eternity.

It was impossible, therefore, to dissociate the motion that bears us upward, which I have called vertical, from the longitudinal motion that pushes us toward the future. Opposed time currents could not be separated either by a master stroke of sensibility or by a master stroke of intelligence; this had to be left for death to accomplish.

I would like to follow this line of thought from now on, for I believe that it is the only one that can explain the nature of time.

Why is pleasure so rare, so precarious, unless it is that we are not intended to be present in the present? Presence is concealed from us, as the idea of the soul was hidden from Malebranche. Can we claim that we coincide with the present via attention? Immediately one would have to point out that we cannot act without a certain distraction, or, as Bacon said, *tanquam aliud agendo*. There has been much discussion of the mechanisms of inhibition that enable us to function in the present without being hindered by parasitical memories or inopportune impulses. Apart from these psychic mechanisms, however, are there other and more spiritual predispositions that compel us to pluck only the flower from each present moment?

Everything in life takes place as if nature were seeking at one and the same time to maintain us in the present and to deprive us of it. When we sense the density of it by chance, as we do when we are in pain or are bored or are full of expectation, what we are sensing is a false profundity caused by a physical disturbance. This is why we always find it easier to endure a current difficulty than one we only envisage. When we forge an idea of some eventual trouble, we endow it with a fictitious presence that it will not possess when it has become real; the present comports a kind of grace of detachment and distraction; if we deliberately relax and surrender to it, it will facilitate action and even help us to endure pain. But for all that, this mental or physical presence does not give us the sensation of presence; it is like a spiritual automatism that we would upset were we to concentrate on its constituent parts; presence must be sustained in relation to the whole of the action.

People who have behaved heroically in the face of

sudden danger report that once the decision to act has been made, the succeeding psychological state in which they carried out the action could not be interrupted. Public speakers are familiar with this absence of any sense of the present as they are speaking, and the more easily their words flow, the more true this is. People (Montaigne, for one) who have lived in fearful anticipation of death have been surprised at how simple it is to die. William James was in San Francisco during the famous earthquake, and he was astonished to discover how well prepared he was to meet catastrophe; anxiety, he remarked, requires distance.

Here is one notable difference between what the present means for man and for an animal. When the animal suffers or when he is enjoying himself, the one or the other sensation occupies him entirely. We could even speculate whether his physical pleasures do not resemble suffering more than ours do, since, to judge from external evidence, they consist much more in being freed of some discomfort than of experiencing something agreeable. (Hence the pleasure of the man who has let himself be animalized seems much more like the suspension of an unbearable languor than like pleasure.) The classification of sensations in two categories according to whether they bring us pain or pleasure is scarcely to be challenged. But we should never forget that there is a neutral, indifferent foundation, a kind of diffused fundament that envelops the senses and can cause agreeable or disagreeable sensations according to how we interpret them. In any event, animal sensitivity is imposed on all living creatures and encloses them within the present. Human sensibility is in part the

86

product of the mind, which works with only one material in the present. Pleasure and pain visit us whether or not we will them, but qualitatively they are not beyond our control; each of us has the kind of joy and pain that he deserves.

Furthermore, when man has evolved and has mastered the brutality of his instincts, sensual pleasure has no real savor unless he can extend and transform it, and this he can do only with his mind. Pleasure confined to the present moment would be simply a thrill. For pleasure to exist, we must know how to keep our distance, as it were. Each of the senses is cultivated according to an individual aesthetic, but the basis of this art is in all cases the same. It consists in deferring and in recapturing sensation rather than in instantly experiencing it. In this area, talent consists in arousing a concert of memories, feelings, and thoughts to accompany sensation. The Epicurean generally draws on the senses to produce such an accompaniment, because the ideas and images he summons to support sensation are subordinated to a primitive pleasure. It is quite different with the spiritual man. He is reluctant even to speak of pleasure; he would rather let it be understood that he is quite detached from pleasure, yet he, too, possesses an art to stimulate it. Here the senses play the role of servants, and his mind draws on them for only a mild excitation. He finds a sustaining rhythm in them as well as a symbol of the processes of the mind. We observe this in poetry or in church liturgy—and, very simply, in speech.

In these circumstances, how well equipped are we to grasp the meaning of the present moment? We know what we have wanted to do only when we have already done it. We do not grasp our real intention until we can examine our conscience, and at the moment we act this is impossible. The present is always the most confused moment in history because we never know whether the current event is germ or dust.

For this reason, we do ourselves a disservice to boast of our modernity. The press—a more significant name than we generally realize—would not survive if it did not feed on the illusion of crises, but for the reader a measure of skepticism is the better part of wisdom. Children have a penchant for endless mishaps, and education consists in making them unresponsive to everything that is not noble. The perfect man would not recognize that something had happened were that something not a mistake. In any event, he would not feel hindered by failure, for in his eyes there would be no defeat or victory except at the end of time. The whole of life, not its separate parts, is susceptible of qualification. We have a feeling of triumph or of disaster when our perception is concentrated on a detail and prevails over reason, which apprehends the sequential linking of events. The more fully awareness functions, the more emotion diminishes, and here again we can learn from the child's example.

The child is enchanted only by surprises. For the grown man (except in games or play, when he recaptures a kind of childhood because these activities are not work) enchantment is priceless only if it is assimilated with daily life; the miraculous is not cherished unless it is dissimulated in the ordinary course of events, in which case it becomes mystery. Novelists realize this; when they want to lead us into a world of fantasy, they know they must do it imperceptibly; we must find ourselves in the midst of the unreal suddenly, when it is too late to be on our guard. This is also the technique of seduction.

Many people believe that we are in contact with the material world that surrounds us via the senses. Nothing is more debatable. Sight does not so much bring the observer and the thing seen closer as it establishes the fact of distance between them, and most of the objects it does bring within our view are forever beyond reach of the other senses. This accounts for the fact that once upon a time the sky was worshiped—it looked so unlike earth. A touch sensation must, in most cases, be transformed into a feeling that in turn has more to do with knowledge than with pleasure. Normal life is a day-in, day-out affair. Tragedy is a phenomenon of retrospection. True, the novel and the theater paint a very different picture of life for us, but it is the law of both to transpose the even tenor of events into crises. Nature, it seems, has sought to deprive us everywhere of the sensation of being: to sense a thing is not to know it. What we term ennui or *déjà vu* or monotony is often the veil that must interpose itself in the course of our activities between the mind and life to prevent their absorbing and ruining each other. The impression of banality plays a role analagous to that of decency.

It is very hard to give a definition of decency that takes into account its full meaning. But whatever its origin and its mechanism, must it not serve some purpose? I never tire of admiring the precautions nature takes to preserve new and still maturing life. With every seed she sows, she must solve a problem: the new creature must be protected against all infectious contacts and yet at the same time it must take from and be nourished by its surroundings. Fur, bark, skin, everything that is tegument, everything that envelopes, fulfills this function in animal and vegetal life. Our sense of decency has a comparable function. It protects the mind from being beguiled by its constant companionship with the life of the senses. Modesty, which is a private or intimate form of decency, acts in a similar way to protect our moral nature from the giddiness that might arise from self-contemplation or, were it assailed by praise, from self-satisfaction. A warning flush mantles our cheek in both instances, perhaps because the danger is analagous.

Decency, then, weaves a translucent screen between the senses and the mind, between action and its success. Contact is permitted on condition that it be immediately forgotten once it has served its purpose, which is to stimulate the mind and nourish the heart. And so we may say that decency leaves a modicum of non-experience within experience itself.

When decency is further spiritualized, we call it intimacy. When confidence is limited to a chosen few and is tempered further by silence or symbolic expression, intimacy has once again performed its task, which is to apply along the edges, softening but not effacing the contours, the *sfumata* that Leonardo said was the final work of the painter.

Furthermore, our attention is not really directed to the present. When attention does focus on something current, it arouses a sense of uneasiness and imbalance.

For example, if we listen to ourselves speak, we run the risk of stuttering. Attention should be directed to the global action or, better still, only to the future, for which reason waiting is the quintessence of attention). We are interested only in what is not, or rather, in the presence of that which is not yet ensconced in what is, and that we call our "sense" of it. We are interested in the promise, in what must be divined rather than known. The artist turns away from the work he has finished. Introspection is difficult not only because it divides us into two beings but also because we are designed to look forward even more than inward; nature wants us to be aware of but not completely present to ourselves.

Thus the posture of pleasure does not fit the normal structure of being; in order to enjoy (that is, to summon up an image of the temporal-eternal), the spring of sensation must be released. It is avidity, the idea of the eternal in the present, the form of this matter, that gives pleasure its genuine falseness.

However, the various conditions of nonhaving to which dissociation leads also seem not to square with human reality.

Let me go back to the end of the last chapter and develop further what I dealt with only briefly there. I said that the ultimate effect of dissociation is to divide the individual into two parts and that one part is reality, the other appearance. Actually, appearance exists only for the mind; it is the impression of inevitable and false presence that one is forced to deny but that nonetheless exists. For Plotinus, it is a plane of time; for

Spinoza, duration and imagination; for Gide, the plane where good and evil are separated. The second part, reality, is the one to which we rise by a total spiritual change, and it is urgent that we do attain this plane because we have emptied the first of its substance. We are now lifted to the plane of ecstasy, of supreme knowledge, of the eternal moment. But this state, although by right a normal one, is achieved only occasionally, its essence is given to us only accidentally, we do not attain to what we insist must be; accordingly, we are forced to substitute nonbeing for being, and desire for possession.

Theories of noetic salvation loftily claim to have broken free of the illusion of matter, the hallucination of appearance, and any vulgar belief in gods and things. They accuse their opponents of idolatry, as we see with Spinoza, but the reproach they level at others applies more particularly to themselves. In actuality, they do not get what they promise; the eternity they presuppose remains a matter of faith. They reject hope, yet they themselves are the incarnation of hope. They deny matter and disdain worship, yet they make a worshipful thing of the movement of the mind.

For Plotinus it is ἔφεσις; for Spinoza it is "intellectual love" (love that is postulated rather than felt, as in Christian grace); and for Gide it is the extended, empty condition that, contrasting it explicitly with love, he defined as fervor.

Any such doctrine must ultimately develop into some kind of mysticism, or rather it actually is a mystique, and not surprisingly all the values proper to religion have been transposed into it. If eternity were attained, mystical life would be finished, since no one can continue to aspire after what he already possesses. But although eternity may be attained in principle, it is never achieved in fact, and so we must live in a state of still incomplete possession, of unsatisfied desire, of unproved insight, of tested love, and, if I may put it so, of absent presence. Are not these the very motions of the religious spirit? However, since the object of faith has disappeared, attitudes and ideas that one wants to preserve despite everything will have to be clothed in a new meaning.

We detected this transposition in Plotinus, who took certain ideas fundamental to religious life—the fall, memory, history, justice, afterlife—and integrated them into a universe where time has no place. The transposition is more studied and more subtle with Spinoza. Not time but the idea of creation is eliminated, as is the idea of a free, gratuitous, and prevenient love. The love that binds God to the soul and the soul to God can only be a love of God for Himself. Conversion cannot involve an initiative on the part of the soul; it is simply an inner rebirth that takes place when we have disengaged ourselves from the snare of appearances; it is our awareness of our salvation. We do not pray; we make a continuing effort intellectually to demonstrate the reality of necessity. With Gide, the effort is made at the level of feeling, and the transposition is in Protestant terms. We renounce salvation but retain the kindred idea of initiation, renounce anxiety but retain agitation, renounce torment but retain perplexity; renounce sacrifice but keep effort, even heroic effort; renounce surrender but retain deprivation. We give up self-detachment but keep availability, give up hope but keep aspiration, give up charity but keep compassion, give up the love of God but keep desire and adoration, give up love but keep the fervor of love.

If belief, if adherence to a subsisting truth, is called "form" and if, on the other hand, we term "matter" the affective state of the believer, then we can say that these different doctrines retain the substance but reject the form of religious life. Yet would such a separation not be the supreme temptation to the mind? This substance is all that is immediate: it is the "pure love" of quietism, it is tranquillity, it is the feeling that, the young Goethe said, is everything—"*Gefühl ist alles.*" How could the mind bring itself to believe that this is form, this matter? Would the mind not be tempted to separate ecstasy, idea, and feeling from their substrata and make of them the supreme value, which it would be absurd and futile to seek to surpass?

According to the positions we have outlined, eternity is equivalent either to a vow that cannot be satisfied or to a sublime experience that can never be achieved. Yet in the first instance are we not the dupe of language? What does it mean to search or to progress, if one knows in advance that the goal of action is the action itself? Is what we term spiritual life here not rather a subtle transposition of the physical order where action is sufficient unto itself?

Does the fact that states of nonhaving exist prove that the subject reverts to its first principle? If it is impossible to experience what is the very condition of experience, then it is in some kind of nothingness that one will have the greatest chance of coinciding with plenitude. How are we to distinguish between quietude and negligence, or between nonbeing which is the opposite of being and nonbeing which is simply nonbeing? In both cases, we are brought to the same conclusion: the only normal and intelligible types of experience are the two that are to be found in ordinary time only by accident, assuming that they are not illusory; the one identifies us with nonbeing, and the other with the infinite motion of divine life.

Let us assume that these systems of salvation can indeed provide what they promise. Let us assume that for one instant eternity is given to us not as an article of faith but as a possession; let us assume that we know eternity not in the shadowy form of a thing we lack or desire but as something which is self-evidently ours. Spinoza used to say that we do experience eternity. Taken at full face value, is this not a self-contradiction? If we were thus deified, all life would be a vision, joy would become beatitude; human speech would be a cry of thanksgiving or, rather, an eternal amen. Philosophy, the function of which is to take the place of this integral experience by letting us perceive it intuitively or await it with composure, would have lost all purpose. What could we seek for in the midst of plenitude? Search would be conceivable only in the form of a joyful discovery of what we already possess.

What meaning could we give to those moments in life that are filled with what we call desire or pain or hope and that are the fabric of all our days? If by nature and by right we are in eternity, how should we account for the continuing bondage of the passions? How are we to understand that thought is never possession but only progress toward it—a progress that is forever incomplete and never without setbacks? Or that liberty is never more than liberation and not without its lapses? How explain that thinking natures are grafted onto animal life and condemned to the cycle of birth, childhood, declining power, old age, and death? And if a man resigns himself to not thinking about all these things, whether he dismisses them as irrelevant or settles for living in some brave, logical dream, what communication can he maintain with himself?

The point has often been made that the most divergent doctrines are also closely connected; they engender each other incessantly, as if they shared some postulate so profound that it defies analysis. Excessive claims to purity or to disinterestedness have been known to lead to the practice of quite the opposite. For example, the authorities responsible for policing public morality were hostile to the Cathari and the Neo-Manichean mystiques of the late Middle Ages. The lives of quietists like Molinos and Mme. Guyon are also most instructive in this regard. The life of Epicurus, too. The masters are immaculate; the snows melt away with the first generation of disciples. But sweeping maxims are not enough; the famous saying that "He who plays the angel plays the fool" has probably never been properly understood. Generally we assume it means pride will be chastened or that an impulse too long suppressed will be released. While there may be more or less truth in these interpretations, they screen out a deeper meaning. Contamination, I believe, leads almost inevitably to dissociation. We indulge too often in the notion that the dissolute man can be satisfied by or, as it were, submerged in sensation. That was possible, perhaps, when mankind was primitive and barbarous. But what happens most often is that the sensory life becomes dissociated: it sustains one part of existence in the material world but, by way of compensation, is able to lift another part up into the ideal world. When St. Augustine described himself as being most alienated, such separation took place, for never did he burn with a nobler passion for pure truth, *liquidam veritatem,* than when he was held captive by the world of the senses.

Remember, too, what we observed in Gide when he speaks of love: he carries dissociation to the extreme; he divides the act of love into practices that are, on the one hand, somewhat more sordid than usual, but he lifts another part to heights somewhat more elevated than is ordinarily expected, and disregards man's habitual level of behavior, of which he disapproves, as the Cathari did earlier out of their desire for purity.

We are not too far removed, then, from life that unfolds on two planes, one of which is eternal aspiration after pure love, and the other temporal and sensual pleasure. There is a closer relationship than we may think between the man who seeks salvation through contemplation of eternity and the man who places the pleasure of the moment above all else. The best and wisest of men is not so absorbed in eternity that he is above indulging in ignoble gratification whether out of weariness or curiosity or perhaps a kind of contempt. Conversely, we see how, out of nostalgia or the wish to restore his self-esteem or the desire for a different kind of experience, the man of pleasure becomes capable of a nobility all the more admirable because voluptuousness has taught him never to treat his own nature with respect.

What is the opposite attitude to what we have called pleasure? We now know in what direction to look for it. We found that the search for pleasure did not err in wanting eternity and temporality at once, for we know that the two must not be separated. The error comes from the fact that pleasure tends to amalgamate what we will deem potentially separable but never dissociate in fact.

Pleasure is not granted to us in time, because it requires a kind of halting of time. If pleasure absorbs time or if it becomes too acute, then pleasure and time are at variance. If pleasure is to be pure, it must be seized as it flowers and passes, but it cannot be held or expected to return. Aristotle was right to compare the actions of pleasure and thought. Both occur instantly, like a vision; they are not, they are, they are not. The Epicureans also understood that one must be abstracted in the midst of sensory satisfaction, and we see the same in Montaigne, La Fontaine, and even Renan. Now we are very far from but at the same time very near to true wisdom. Far from, because the lover of pleasure secretly believes that life has no consistence but is a dream or a mockery, whereas it does possess substance. Very near, because the dilettante has understood the wise man's advice, which is not to enjoy but to use.

The word "usage" has acquired so many meanings in its long history that I should like to suggest a fresh one. The Romans, more precise than the Greeks, contrasted *uti* to *frui: Kannibal, quum victoria posset uti,*

On the Structure of Time

frui maluit. Seneca and Cicero said that one is using
the good when one enjoys light. The Romans called
bread *"frumentum."* Pleasure was an activity that pene-
trated to the innermost part of being. It becomes egois-
tic only by a deviation, for its primary nature is such
that it connects us with, or puts us in communion with,
what is true and substantial. We are able to rejoice in
the happiness of others, although we cannot make use
of that happiness for ourselves, precisely because it supplies a satisfaction that, in effect, changes us into another person. (This is the true meaning of "congratu-
late"—to wish another abundant joy.)

Therefore, we should reserve the word "enjoy" to
spiritual possession, as the medievalists did, and Spinoza
after them. That is to say, we should "use" material
things but "enjoy" only spiritual things. Confusion
arises when these attitudes are reversed; the upright
man enjoys God, and this is the basis of his glory; the
lesser soul uses God, whence his deceptive joys and his
torment.

The distinction raises the meaning of "enjoy" to a
very high level—higher, perhaps, than the average man
has been able to live up to. Since it is he who creates
language by speaking it, the meaning of "enjoy" has
declined. The "enjoyment" proper to the spirit did not
seem valuable enough to deserve a word all to itself.
Sensory satisfaction also came to be designated as "enjoyment." The idea of "use" declined in its turn;
"practical" has become synonymous with "convenient"
and "comfortable." Similarly, the meaning of "custom"
has come so close to that of "routine" that for the rela-

tively unschooled speaker "routine" and "custom" are one.

For our purposes, "use" will denote any action that is taken with regard to temporal matter and that gives it meaning. If the material thing is an object already fashioned and converted into a tool, to use it makes it serve our ends. Now let us eliminate one element—the material—and keep only the other, the temporal, namely, eliminate matter and retain time. We will say that use is a property of time, or that time is the place of usage. Man cannot be coincidental with time; consequently, he cannot enjoy the things it offers him. Enjoyment would require that time stop and that it make a transition into its profound dimension; however, we have seen that this transition is as impossible for the senses as it is for the mind. Yet if the lower pleasures cannot be sustained, and the higher ones scarcely more so, we can still give meaning to temporal matter.

From this point of view, use is superior to privation (nonuse) because it enables us to fulfill our human function. Spiritual men have conceived of the use of material possessions in this light. If they have seemed to criticize usage at times, it is because man's faults weaken him so that it is difficult for him to use without enjoying. But once the spirit has been purified and has passed out of darkness, its false capacity for pleasure has been reduced; it can then resume the use of material things that it had relinquished as a precaution. Only then is the spirit using the true essence of things, and so restores sensory experience to its proper purpose.

Ascetics have condemned human sensibility because it is the prime example of ambiguous powers; faced with the difficulty of how to use it rightly, they preferred to take no risks. However, asceticism runs another danger, namely, to forget that sensibility has its necessary office for which there is no substitute. When it contributes resonance to emotion, measure to thought, and to desire its harmonics, when it puts us in intimate accord with others, when it transforms ideas into subtle movements of the blood and nerves that an inner sense can perceive, when it permits us to predict accord and discord even before we can understand them—then sensibility is irreplaceable. We see this in people who have impaired it through clumsiness or rigidity, and even more clearly in people who have allowed their sensibility to become muddied and, finding themselves unable to reclarify it, can only repress it.

Some people may believe that a man is more a man for renouncing sensibility, but actually he then resembles a spiritual automaton rather than a pure spirit.

Use establishes us in the present in a very different way from pleasure. Pleasure does not require an act of will. But usage requires us to narrow the field of our awareness and to sacrifice any excess of present life. We must have a mechanism at our disposal that allows us to adapt to the smallest opportunities time offers. Simply to let time flow by in an empty, monotonous flux is *not* to use it. The brain is the tool of usage; so is the hand, with its thumb that by pressing against the fingers allows us to pick up and to relinquish things. But our use of things is never more perfect than when we employ a tool small enough to allow the body a point of application; the more specific the point is, the more perfect the application. The lever is such a tool, and so are all instruments that need a fulcrum.

Writing offers a good illustration. A line is the undulation of a series of points. Of all manual skills, writing helps us most to master ourselves, because it requires us to make point after point advance across a blank space, and to follow contours that are all the more complex for being governed by convention. (Handwriting always reveals any mental disturbance.) A footstep, a heartbeat, the phases of the moon, and the cycle of the seasons all are signposts that enable us to reckon our own movements. Instruments—witness the clock—need only imitate and extend their action. Thus the tool not only affects space; it also discomposes time. It scans and moderates the activity of the mind and disciplines it for effective use.

Usage is frugal and must accordingly dispense with two types of responses: one, having to do with the future, is delay; the other, having to do with the past, is regret.

People who prefer to act only on the spur of the moment are prone to excessive haste, which results, as we have seen, from loading the moment with a greater charge of life than it can support. Let us stay within the limits that nature has prescribed and try to restore action to the proportions it should have. Since there is always a margin between what the present is asked to accommodate and what it can accommodate, some part of the action must be deferred. It is the wise man who can put something off until tomorrow. (Laziness is simply a deformation of this wisdom.) Justice has found no way to secure even a precarious peace except to fragment a trial into so and so many distinct parties, determine adjournments, reserve the right of appeal—not only to wear down the contending parties but also to control a necessary period of delay (much as tragedy disciplines expectation). Delay represents victory for the weak, and since the weak man is generally in the right, delay will be, as it were, his salvation. The violent man cannot wait.

The various ways of contriving delay are easily perverted, of course. Delay may deteriorate into a lag; an attitude of waiting may become indolence and ultimately resignation; putting something off until tomorrow may lay the groundwork for permanent postponement; a court adjournment can become a parody of justice. However, if we weigh the advantages against the disadvantages, delay still comes out the better; the man who can defer action, who can put off until tomorrow what he cannot do today, has already conquered time.

Parallel to delay we have regret, which is delay in the past. Of necessity, usage leads to regret. Perhaps too little attention has been paid to the fact that there is a normal regret, a regret that is indispensable to our human equilibrium. It is related to the true essence of time, and derives from our inability to realize our desires to the full. It is allied to the disappearance of what has just been. A temporal creature cannot escape feeling that life is fugitive. This is entirely natural, and the elaborations on the transiency of life which are so dear to poets seem rather futile. Why be astonished by a condition that is necessary to life as we know it?

Regret does not derive only from the incomplete nature of the present, however. It comes also from the fact that we cannot act on what has been. Even in memory we cannot add that supplementary quantity of life to a past experience that would make it perfect in recollection. True, in the moral sphere correction is possible at any moment, and this kind of hope directed toward the past we call repentance. Yet repentance can erase error only if error is well identified; it can also correct a habit of intention. But aside from this personal and, as it were, accidental difficulty, another problem exists which is connected with the constitution of time and which regret acts to offset. The charge of life that the present cannot accommodate is not only deflected—that is, delayed; it is also dissolved in regret. What we did not do, what we were unable to do—whether because it was beyond our range or our capacity or because it called for a supply of energy that we can muster only for emergencies—we still cling to the notion that we could or should have done it, and that we would have done it had we summoned up just a little extra strength. Now, this discrepancy between the past action as it actually occurred and the action as we visualize it in retrospect gives rise to a mild emotion that we cannot call either sorrow or joy but that is inseparable from our use of time. A person who did not experience regret of this kind would be abnormal and any self-satisfaction he felt would certainly be suspect. In some people, of course, this particular function of regret is weakened and deadened by vanity or by an almost physical euphoria or by a deteriorated ideal, but the normal conscience cannot help reverting to the past in order to improve and perfect it; it repossesses the past, just as via the plan or intention, it reaches into the future. These two contrary actions, which constitute human time, give the present its double dimension. Let me also point out that it is in and through them that I am conscious of being free; were freedom confined to the pure present, it would be in danger of being confused with spontaneous *élan*. We should note that those who have denied freedom have also felt constrained to dismiss human design and especially regret as illusory.

I should like now to discuss a new characteristic of this "usage time," which is oriented much more toward pain than toward pleasure. Whenever we confront several ideas at once, we customarily treat one in function of the other. For example, time has suffered a long association with space, and likewise pain with pleasure. By nature, pleasure is instantaneous; it proceeds by a series of jerks; it is bound to the sensory organs and their responses; it requires variety to multiply its hidden possibilities, for man must not wear out his pleasures. Accordingly, to maintain oneself in a state of unalloyed pleasure calls for great adroitness. The case is quite different with pain and grief, which operate in a much deeper zone of being. (Language points to this via the double meaning of "endure"—namely, to survive or to last, and to bear or to support.) "That will last me for a while" generally denotes a degree of discomfort resulting from too much of something; before Mr. Bergson set about annexing duration to pleasure, both endurance and patience were ensconced by the side of pain. If we question relatively simple people, we find that they believe pain is the normal, the common stuff of life; pleasure can work its embroideries on this material, but pain remains the fabric of life, and the thing that enables us to snatch something of substance from it.

This almost imperceptible suffering is bound up with contentment; it is a neighbor to peace. It is a mistake ever to divorce grief and joy. They are, after all, two aspects, two attributes of a single and more profound reality. A most admirable characteristic of our Western civilization, one that does it perhaps its highest honor, is our acceptance of hard labor and our having sought the source of human dignity in work.

Sometimes one wonders where to draw the line between Horace's *carpe diem* and the Bible's *Sufficit diei malitia sua*. Both recommend that we postpone thought and preoccupation until tomorrow and concentrate on the present moment. But whereas the pagan maxim orients us toward a search for pleasure, the second counsels us to endure our travail. The first, were we to push it to its logical conclusion, would have us apply our intelligence to refining and multiplying the savor of the moment. The second leads to an attitude of love whereby the tree may bear fruit in its season.

It is a matter of common observation that joy and pleasure do not ordinarily make for a compassionate heart. Indeed, quite the opposite; if anything enables us to communicate with our fellow creatures it is the fact that we have laid ourselves open to pain. Compassion has the effect of augmenting the density of any given moment because several consciousnesses share in it. This is why suffering is so effective in expanding true sensibility, and pleasure so much less so. The pleasure-seeking man, we know, grows less and less sensitive to pleasure; he is forced to vary and complicate it in order to feel and respond to it. Perhaps worse still, he watches his capacity for compassion shrink.

Just as illness makes us aware of the ailing organ, so unhappiness develops a new awareness in us; it sensitizes us to modes of experience and knowledge that we would never have discovered without our having suffered. The sensitive person is one who has suffered more than others and in circumstances that should not have led to pain, and thereby he lends, so to speak, the capacity to feel to zones of the soul that had been until then untouched and virginal but also incapable of communion; he has equipped himself with new senses, let us say with new antennae, that enable him to apprehend the invisible. Thus privation has an unsuspected power and, as a means of knowledge, is often superior to possession. Every deprivation binds us to the thing desired and causes a kind of pain, immanent in life and in time, that is essential to the nourishment of the soul.

There are other characteristics of usage that I might mention. I am thinking of some that are less apparent, that are not associated with the trilogy of past, present, and future, that are components of it but that have more to do with the modes and effects of usage than with usage itself. In particular, I am thinking of one characteristic whereby use is subject to the law of suspension and resumption. The men of antiquity saw perfectly well that repetition is essential to the flow of time, and they defined time from this point of view. Aristotle called time the number of motion; that is to say, for him time was the characteristic of motion that makes it calculable in terms of the repetition of its phases. The early philosophers remained on the level of sensory observation. Modern science has shown that they were immensely more right than they supposed, although the world is quite otherwise than they thought.

The law of rhythm and of repetition is to be observed in nature at every stage. In creating life, nature had to submit it to the law of age, that is, to the pattern of life, growth, and death. But within these limits nature could have restricted the role of rhythmic movement. It seems that, on the contrary, she wished to augment that role and to make it more complex. The higher the form of life, the more varied and complementary the rhythms to which the creature is subject. It is possible to conceive of a type of organism that would assimilate the sustenance necessary to maintain it without benefit of any rhythmic mechanism—for example, the way photographic film fixes light. Such a mode of organization prevails in the vegetable kingdom, a plant absorbing solar carbon, but it is not to be found in the higher organisms. A higher organism does not take nourishment once in its lifetime or once a year or once a day or once every hour or minute. The animal machine must dip constantly into its environment for replenishment; therefore, it needs mechanisms that can both take and reject, and since energy is dissipated in work, they must be capable not only of exertion but of rest; hence the pattern of a moment's returning and resting, a stopping and a beginning anew.

This is why every living creature exhibits a variety of pulsating, jerking, and wavelike patterns, which are the more marked the more complex the organism.

Joan of Arc testified that she used to hear her voices more distinctly in the evening and in the measured ringing of the bells. Our blurred perception of the rhythms of the body and of the universe supplies some such service to the mind. In 1815, Maine de Biran noted how "each season has not only its kind and order of appropriate external sensations but also a fundamental way of feeling about existence that is analagous to that season and recurs uniformly with its return."

Cosmic rhythms or inner rhythms are only accidentally apparent to our consciousness. But one rhythm is imposed on life that no study of human time can neglect. Think for a moment of the alternation of days and nights; this alternation could be a simple fact of observation or a convenient frame if our nervous system did not find in it the release of sleep, which the mind puts to marvelous use. Astronomical and biological rhythms help man to master himself and to mature. An animal sleeps, but does it fall asleep? It is awake, but does it wake up? Probably its consciousness is always drowsing. Can we conceive of a sleeping within sleep or of a dream within a dream? Only consciousness of the human type can experience this interruption which is the prerequisite rather than the effect of a decision. The state of being awake, which is common to man and beast, is the prerequisite for awakening, of which only man is capable.

We are all familiar with the experience of starting up, knowing that we are but not knowing too clearly where or even who we are. Then suddenly the lines of communication are restored: we are awake. Whereupon we must consent to begin again and to continue, to resume

109

and to pursue—a double assent that allows us to grasp the essence of willing. Actually, we do not pick up our life and work exactly where we left off the evening before. While the nervous system is relaxing and dissipating congestions and fatigue, the ego is restored to itself, and rediscovers its potency and unity. But night provides for even more than this peaceful unification of the whole being, when the several parts are restored to their principle. When an individual submits to sleep, he expects that the new day will bring renewed possibilities, new bases for new departures. Nothing is more painful than to awaken and find things as they were. One should never pick up exactly where one left off. Tomorrow does not follow on tonight but rather continues it. Something has been accomplished in the interval—one has grown. And this is the last characteristic I want to discuss in my definition of usage.

The external rhythm to which we are subject immerses us in nature and in life in order to set reverberating in us, as in a particularly sensitive monad, all the pulsations of the universe. However, this is not its only function. It also, I think, regulates and balances the movement of the mind. I spoke earlier of how certain things act as protective envelopes and shield the mind from contact with sensual pleasure. Rhythm has an analogous role, to the extent that it prevents the mind from overexertion and fatigue; it imposes a salutary halt. If there were no such rhythm or no fatigue that alerts us to it, would either effort or pleasure ever be halted? Would not activity of all kinds be pushed to the point of self-destruction?

But within this first rhythm which is imposed on the

soul from without via the bonds that the spirit establishes with the physical universe, there is another,
an interior rhythm of a slower periodicity; actually, it
is not so much rhythm as development. If we picture to
ourselves how an idea grows in consciousness, we notice
that it passes through several describable phases. At
first, there is a tentative searching, which helps the idea
disengage itself from earlier formulations that, in a
sense, may be said to choose it rather than being chosen
by it. In any case, even if they were to envelop it, they
would remain alien to it. Then the idea flowers, it
creates its own appropriate forms of expression and
diversifies them with perfect ease, as if to prove its inde-
pendence of them. It stumbles into difficulties and
triumphs over them by a process of correction and clari-
fication. Even when the idea is mature, the search continues. Finally, in its last phase, which is fulfillment,
the idea achieves a state in which it is no longer susceptible of variation but in which there is life nonetheless.
Then we observe something grow that had not seemed
capable of further growth. From phase to phase, a
spiritual reality that seemed to have attained its perfection and term is enriched by an unknown quality
because as it passes from a lower to a higher plane it can
assimilate a fresh plenitude. This is what happens in
happy love relationships; as the lovers grow old, earlier
emotional conflicts and reciprocities disappear even
from memory; feeling has been delivered from the tran-
sience of time, released from life's servitudes, and has
become a unified emotion that now simply endures.
Like is added to like; there is no interval of desire or
regret but a kind of sweetness of life, and one cannot

111

say whether it is habit constantly deferred or progress perpetually repeating itself.

Thus, at the very core of usage, a kind of fourth temporal dimension is to be found. While it does not suspend the flow of time, it places us beyond change. If there is any change, it is rather in the nature of maturation, which is a sign of spiritual life, or more exactly, is the sign that our spiritual nature is escaping from the vital nature in which it has been enveloped until now. Maturation never shows to better advantage than in old age, where the process continues but is never completed.

Both old age and early childhood are metaphysical ages because of the disproportion in each between physical weakness and mental energy. While physical and mental development parallel each other so closely that they often seem to merge, they are oriented in different directions. In the case of an old person, bodily growth has stopped years since and changed into a more or less general involution, but the mind continues to pursue a forward course. Whereas physical duration is essentially parabolic—involving as it does growth, peak, and decline—spiritual duration permits continual spiritual growth through a silent enhancement of quality.

For this reason, we often find that the novelists throw more light on the question of human time than do the philosophers. Once we divest time of any romantic swaddling clothes (English and Russian fiction is often less exclusively concerned than French with love as a theme), we can follow what I have called maturation. We follow it better still in candid autobiographies in which the individual conscience speaks for itself; here, allowances being made for distortion due to a wish to aggrandize or belittle the self, we find the line of development caught in all its purity, or at least in its immediacy, as it was lived by its first and only witness.

The foregoing discussion has clarified why the techniques of the moment are inadequate, whether they try heroically to lift us above the flow of time through a kind of ecstasy or whether they aim to do no more than counsel us on a given immediate action. If the moral standards of the eternal present are artificial now and then, it is because they presuppose a time that is different from the time available to human experience. Nonetheless, I believe that they teach something very true and necessary, without which all usage would be incomplete, namely, that human time cannot exist without an irreversible evolution toward eternity, which subtends it. The error in theories of pseudo eternities is to misconstrue tendency for reality and to mistake what is only a symbol for substance.

Nothing could be more helpful at this point in our analysis than to consider once more the beyond-the-present that we call the future. The future is the most perplexing and obscure element in human time, but it is also the most richly instructive.

Day-to-day action would be impossible if it were to imprison man within a cage that had no exit. Yet the reality of the future is ambiguous. Everything—the structure of our mind, the direction of our wishes, and what I shall call the precipitancy of our being—inclines us to invest the future with consistence. While the physical plant we call our body makes us cling to the present, the soul makes her escape with only a meager image to sustain her. A little reflection will persuade us that the future belongs, as does its image, to that order of realities that act as bait and as opportunity. As things often turn out, it would seem that there must be a divergence between the goal the individual sets himself and what he is really working toward, because in actuality he is following a plan that outstrips his intentions.

This is conspicuously true of love. We have seen completely legitimate marriages in which neither partner has consciously thought of having children. At the moment of marrying, how many people do clearly intend to have offspring? It even seems that nature did not wish the act that unites the human sexes necessarily to be associated with new life. If we examine the mechanism of conception, we discover so many delays and hindrances to the encounter of the two half-cells required for conception that life might seem to have devised a brake on life. And if we envisage the sexual instinct from the point of view of conscience, we discover that its drive is all toward the pleasure of love and that it is quite indifferent to the joy of paternity. The moralists will have to busy themselves with such matters if they want to explain what humanity has to this point been satisfied to use—and rightly so, no doubt—without trying to understand it.

Yet this attitude is no longer possible in an age that is implacably bent on imbuing the instinct with intelligence at the risk of corrupting both. Recently it was proposed that we distinguish between the purpose of marriage—the child—and the end of marriage—the union. However this may be, love between the sexes in the human species is a mystery. Nature seems not to have dared rely on the good will of her one rational creature to insure the continuity of the thinking species. She wins his assent by appealing to his senses, and then lays on him the burden of rearing children, with all its attendant woes, worries, and responsibilities. How strange that to insure man's making the sacrifices which so ennoble him, it has been necessary to bank on his futility. Still, when men want to make other men behave like heroes, they do the same; only yesterday soldiers were lured by baubles of glory, and how many of them gave their lives in return for a ribbon, yet thanks to this ruse the homeland was preserved. Or to take an example from another area: the search for spices led men to undertake perilous voyages, and by way of bonus new continents were discovered. If one has really understood the meaning of "instigation," none of this is surprising. Probably the living, thinking human creature does need a needle to arouse powers that are asleep within him. Nature and society have provided with the greatest care to release this energy, as if the essential thing were to clear the decks, after which action would flourish according to its own laws and, above all, would subordinate itself to the common good.

I have dwelled on these examples at some length because they offer in expressive symbols what is everyone's image of the future. Often, if not always, the future plays the role of bait and needle. It urges us on toward some immediate, small, often artificial and empty benefit. But the action that this image sets in motion infinitely surpasses it in worth and dignity. Basically, what we believe we want is small things: pleasure, success, approval, a little money, comfort or recognition, leisure with respect, *otium cum dignitate*. We make a sustained effort; we engage the whole of our being to attain these ends. It would be wrong to judge us in terms of the incitements that make us act. They are there only to set profound currents of energy and love to flowing. The one danger is that once we are aroused by this bait of the future, we will stop short with the image and construe it as eternal. We are not wrong to fix our attention on the present, to put off until tomorrow, to want what is not yet, to live in the future. The depth that pleasure lends to the present is fallacious; the normal depth of the present is what comes to it from the future. Even if this depth were illusory, it would still sustain action, and in any event it is through the intermediacy of the future that we can commune with the best part of ourselves.

It is remarkable that even the idea of the past comes alive only for those who are in some degree committed to the future. One would assume that the historian is interested in the past as such, and indeed some scholars have this vocation, while some historians are merely curious. But the great historical works have been figurative. They have selected from the past the things that explain the profound significance of the present and throw light on the still obscure image and analogy of the future. If so-called ancient history is more instructive than any other, it is because we see in it the springs of events in a more manageably simple form, which allows us to balance the present and evaluate the unknown.

The future, then, is necessary. But it is always ambiguous and in the mixture it offers us, aspiration toward the eternal via the mediation of the image desired is constantly confused with the *élan* of vital time. This is why eternity appears to us as the future and why the future seems eternal. We have therefore to be objective. The problem is to sort out the desire that impels us toward some future object from the aspiration that lifts us toward the eternal, of which the object is only the sign. Language lacks exact words to help us here, and understandably so, since the office of language is to serve life and for this purpose ambiguity is too useful. If a clear idea assists thought, an obscure idea is already an action. As we have seen, St. Augustine had two terms at his disposal for which equivalents should be found at least in the inner language that each man speaks to himself: *extensio,* or desire that projects us toward the future; and *intensio,* or desire that lifts us toward eternity.

It is a profound distinction that he makes here. However, let us not regret that our own language has not accommodated it, for the separation of these two currents cannot and should not take place. Philosophies that have dissociated them in the hope of eternalizing particular segments or directions of time have erred, for they have never been able to recover unity of being or the savor of life. The fractures that exist for the infinite being who surveys them eternally are not knowable or separable for the finite being who experiences them in time. That is why the flow of time is more rational than the mind itself. Must we, then, not distinguish between the two currents? Yes, indeed, but by one of those acts of intent, the essence of which is never

to resolve the problems that sparked it. These actions
are very important in the moral life; actually, they
constitute the whole of it. The secret of temporal life—
and what makes it difficult—is to distinguish through
intention what it is quite impossible to dissociate by
reflection, to separate only potentially in order to achieve
unity in action.

Having taken these precautions to dispel any equivocation, I would like to suggest some definitions that
may help the mind in its search without leading it
astray. An analysis of hope versus aspiration will be
pertinent.

Hope belongs to the order of desire. It is the direction of our thoughts toward the future, and inasmuch
as the orientation is usually toward temporal goods, it
can scarcely qualify as a virtue. Aspiration, on the other
hand, is what draws us toward the nontemporal. We
hope for good weather tomorrow and we aspire to
behave well tomorrow. In both instances, certainty is
uncertain, but the dosage is different. If we analyze
hope, we find uncertainty dominant because the im-
possibility of foreseeing all contingencies opens the
field to chance. It works otherwise in what we mean
by aspiration.

Bossuet was right to say that aspiration embraces a
kind of desire. But this desire, oriented toward nontemporal good, has a larger share of certainty; at the outer
limit of good intentions, we would find an anticipatory
possession that does not mistake its object. To hope is
to expect that which cannot deceive us, for if the good
exists, it will be. Hope owes its uncertainty, one, to

the fact that time is separated into moments and that we never know whether the next will not be the last, and also to all the fortuitous circumstances that can prevent our wishes coming true. Aspiration owes its uncertainty to the separation that obtains between conviction and clear-sightedness, between demonstration and evidence, and perhaps even between the evidence we remember having had and the evidence we do have. But its uncertainty is even more attributable to the fundamental separation of time and eternity. The uncertainty of aspiration is always accidental. Eliminate the distance or remove the screen, and aspiration will be transformed into possession. In other words, aspiration possesses with confidence while hope is anxious. I do not deny that the two impulses are not adulterated when combined. I say that it is wise to distinguish between them intellectually even if we must watch them intermingle before our very eyes in the normal course of time.

Let us then term aspiration that irreversible motion of time toward eternity, which aims to abstract time from its flow but cannot actually achieve the separation. Without aspiration *age quod ages* would soon secrete an irresistible *cur agam*.

Moral doctrines are concerned with the efficacy of action, and therefore they seek to purge it of useless cogitation and to set before it an unattainable goal. This is strikingly apparent in the Sermon on the Mount, where the invitation to live in the present day and to take advantage of the sufficient evil thereof is accompanied by the call to infinite perfection, even as the Father in Heaven is perfect. The Lord's Prayer also recalls this separation through its two complementary parts. The first expresses a supreme desire of the mind: it asks that God be, or rather that His will continue to be done. But the second part of the prayer is made up of requests—for daily bread, forgiveness of sins, and deliverance from evil.

An analogous rhythm is found in Stoicism, although relatively it is quite inhumane: aspiration has been changed into the assurance of a total wisdom that is possessed in perpetuity; the Stoic conscience dips into this as into a well, and draws up its concern with small, specific duties and with a never perfected refinement of intention, and a forever alert benevolence. (Witness the *Meditations* of Marcus Aurelius, which are more on the order of moral examination than metaphysical speculation.)

Let us note, too, that the therapy of the passions seems to be dual in nature. It offers, first of all, the discipline of retrenchment: the mind submits to the laws of daily life and here we utilize time with all its potential of multiplicity, usury, and obliviousness. As Spinoza saw it, time works for us in that it supplies the opportunity for us to combat passion with rational maxims, to divide passion against itself, and to the extent that we understand its causes, to change it into idea. Yet in itself this would not suffice, if we failed to see that passion has a positive virtue: it contains within it something of the infinite, and infinity cannot be vanquished except by itself—that is to say, by substituting one infinity for another. One cannot master life except by means of a higher life; one triumphs over evil only with good.

120

Anxiety is cured not by serenity but by a greater anxiety which does not suppress the first disquiet but rather out of which comes a peace that surpasses it. Turmoil is cured not by tranquillity but by a superior agitation that arises from some worthy cause. An unworthy love is cured not only by being restrained but through a loftier love that preserves the lesser's heat and purifies it. To put it more generally, there is only one way to cure ourselves of an errant impulse, and that is to preserve its thrust while correcting its *élan*. When we desire something that we could obtain but which is forbidden, the remedy is to desire something we cannot obtain. The dimension of eternal aspiration enables it to correct the illusion of the future rather in the way the idea of geometric space in which objects retain their size corrects the illusions of perspective. Without this, we do one of two things. Either we assume that the future embraces eternity, or we abandon the idea of a future and confine ourselves to timelessness. The future then is doomed, and we believe we have been able to dispense with aspiration.

But if the future is to be sufficiently real and meaningful, and if the mind is not to dismiss all the possibilities it extends to us, then our impression of its stability must not be illusory. We have seen that this stability cannot be achieved in time, for the future moment could be no less fluid than the present moment, and the latter has the merit of its charge of hope. At this point, we have only two solutions before us. One is without undue scrutiny to welcome and shield illusion; the other is to dispel it. (Unless, of course, we dare not make a decision and shuttle back and forth between the two, depending on our mood.) So we may say that the future is connected with eternity but is not to be mistaken for it. The nature of the connection is limited: the future is a preparation for and a symbol of eternity; it is an advance toward it and an image of it—I should like to say "sign," if it is true that a sign can be simultaneously image and means.

Kierkegaard said that when eternity and time come in contact, they do not meet in the present, which would then be eternal, but that eternity appears in the semblance of the future. And thereupon a great miracle takes place, for eternity, lofty and unified as it is, divides and offers itself in the guise of manifold future images. Thus she teaches man to hope. Let me put it this way: the future can help us only if we believe in eternity and if eternity is projected in the future. Only then does the possible become real or at least realizable: then truly there is something to be done.

We have noted that the fallacious idea of the future is useful to men who want to establish something or other in this world. Is this the same as saying that we cannot create anything lasting without the ghost of hope? Not at all, yet illusion is one image of truth. If it is childish to imagine a future that will be utterly perfect, only a slight correction is needed to help us find the paradoxical but true equilibrium: we will be well established in time only if we have a hope of eternity. This has been the inspiring idea of our spiritual teachers, starting with Socrates. Eternity attracted them powerfully, yet they were entirely taken up with time. It would seem that they should have been unconcerned with provisional things, that they should have waited for the veil to be lifted and for the inevitable to come to pass. What folly to work for a better time when one has condemned time to the status of mere transition! How does one sincerely prepare for the future if one believes that the future will be precisely like the past? But they saw no contradiction in this; indeed, it is the idea of eternity that helped them to believe in the future and to aspire that it be better.

The future is the face of eternity. And that is why it is almost hallucinatory. And why it alone attracts our love. But for this love to be pure, we must constantly rectify it by recognizing the motionless, misleading images of future time for what they are and divining the eternal stability that lies behind them.

Here the mind is not entirely efficacious, for the mind dissociates and separates. It must make its voice heard, of course, and remind us of inescapable distinctions, yet the mind must also bow to a superior meaning, which is the true knowledge of the third order: the meaning of human time, the meaning of the human condition, and also the meaning of eternal evolution and preparation.

This would be well defined by the expression "eternity in the present" if we did not know how equivocally the term has come to be used. I prefer to say that the knowledge of the third order is that which is implied in the practice of virtues, especially aspiration and patience, which are the pith of all the others. Or perhaps it is what once upon a time we called wisdom.

Conclusion

THE PROBLEM OF HUMAN TIME and the uses thereof is related to the superior rhythm Plato spoke of, which seeks to determine the structure of being. Taking time in this sense, we observe that it is an amalgam, and we correctly recognize that in it a timeless element and an element of absolute mobility converge. How do these two elements assert themselves in relation to each other?

This was the question we set out to explore. Two solutions are available. One says that the two elements must be inseparable; this solution leads to bastard forms; a kind of unity is achieved but only at the expense of structure. The other solution counsels separating what is by rights separable and isolating the eternal—in other words, dissociating. We have seen how

125

dissociation somehow brackets two natures which it cannot unite. Only the eternal part of us is retained, yet it cannot be maintained; the temporal part is proscribed, yet we cannot withdraw from time.

It remains for us to evaluate the two systems, both of which we reject, and to measure by how much they approach or deviate from reality.

Even when one must reject a given doctrine, it may nonetheless be a useful analytic tool. There is no error that is not in some measure true, for often what we call error is a hasty attempt to single out and act on a distinction that is no doubt genuine enough but that exists only potentially in the unity of being. As an example, let me cite materialism (in the highest sense of the word), which isolates and canonizes the functioning of what Aristotle termed the "material" cause. In the case of man, the physical cause is aroused, directed, and transfigured by a principle that one cannot honestly ignore. But this much being said, all matter then regains its rightful place and price, and the truth of the error is discovered rather in the truth of a purer method than in terms of the error itself.

In this spirit, let us trace the line of truth along which the two opposing views of time meet.

Both are, as I have said, doctrines of temporal salvation; they attempt to place eternity within time—as a result, to deprive death of its significance as metamorphosis and passage. Yet this denial of immortality, or rather of immortality as our imagination represents it, has a purifying effect. The practice of morality and especially of religion could easily incline us to belittle earthly existence. But placing eternity after time does not confer real value on time. It does, of course, give weight to our present life to affirm that it is the place where an option that involves an infinite duration is exercised. But by an almost fatal reversal, it also involves the risk of devaluating time precisely at the moment when it is lent a tragic dignity. Quite possibly the person who is urged to yearn for eternity rather than to take up his abode there forthwith, will come to see time only as a means, as a kind of coin with which he will be able to purchase salvation. Furthermore, since eternity is said to be the fruit of sacrifice, and since it is thus in harmony with pain, the utility of time could come to seem purely negative: time would be offered us only so that we should resist its solicitations; it would attract us only to put us to the test; in a word, it would exist only so that we should deny it that supplement of being that is born of consent.

Does this not lie behind the idea that earthly life is the place of temptation? One respectworthy religious tradition (belonging to the moralists rather than to the saints) has insisted on such self-restraint that all temporal pleasure seems by definition to be censurable. On the other hand, pain and suffering have often been exalted because they seemed to restore to time its true function, which is to be the instrument of human trial. Life is granted us only on tolerance, so it is right to renounce it. The prayer that asks God for help in the good uses of illness has something of this attitude in it; we find it also in Kant and even in Lachelier. If such points of view glorify suffering, they do so for a very profound reason: by making us deny temptation and by sparing us the choice of our afflictions, suffering achieves the self-negation that is essentially our temporal task.

This point of view has understandably driven the friends of human nature to protest. Plato reproached the Gnostics for considering life an evil, which horrified Spinoza, too. The doctrines that establish eternity in time do enable us to resist contempt for temporal things. They are antidotes to what I have called morose pseudo eternalizations.

Human life cannot be considered simply as a means to an end. In one sense, life is truly its own end. Ecclesiastes proclaimed this, and so, with somewhat more optimism, did Epicurus. Can a conscious and especially a moral being make use of another being as a tool or means to an end, even if the other were an inanimate thing? Can he use his own present life as if it were only an instrument with which to prepare for another life?

Such a notion seems to me incompatible with the very idea of creation. That the first being freely calls into existence a universe which is the embodiment of his thought is a necessary conclusion in order to explain the integral datum; that God may stop sustaining this second life and thereby annihilate it is, for the all-powerful, always within the realm of the possible; in any event, nothing expresses the concept of divine freedom better than the threat of an interruption in creation. Yet it would be most difficult to reconcile a suspension of this kind with infinite wisdom and reason. How would we explain that eternal creative love could stop willing what it had once willed? Would time be worthy of his having created it if there were not somewhere in its essence—I would say even in its flow—some substance that can always be preserved?

Conclusion

We could reach the same conclusion by a slightly different path, simply by reflecting on the dignity that clothes the tiniest particle of time which enters our consciousness. By the mere fact that it is known, it becomes susceptible of perfection and even of beauty. The first condition of beauty is that it correspond to and be sufficient unto itself. Were we accomplished artists and if daily time were the material we worked with, we would be able to find this beauty in the smallest instants of our lives. We have a presentiment of this when we think of our childhood, that long-ago time when we still had no personal history because nothing had yet happened—or, rather, because everything that happened was to us an event. Furthermore, since human beings belong to the moral order, intention, when it is genuine, is sufficient to imbue the attempt and even failure with the perfection of the completed action. For that matter, in one sense perfection is a quantity that may be instantly acquired. Then why wait for it so long? What ease of habit and virtue we possess is achieved only through practice and could not survive without regular exercise. To qualify as a master, we must remain an apprentice to the end of our days and secretly relearn what we teach and believe we know. But there is a royal and holy perfection in which we can only establish ourselves instantly, for it is an illusion to believe that the following moment would be more propitious in helping us. Effort, much less spontaneous *élan*, has no precondition: nothing prepares the way for love but love itself. What is needed, then, is some way to raise the instant to the ideal plane, or even to draw perfection down and establish it in the present moment.

When the purity of intention is such that this facile miracle has been wrought, we feel sure that the moment to follow and even eternity itself will offer no more.

This is why haste is ill-advised, why a certain slowness of pace is needed in order to savor life. How Goethe has been taken to task for his eternal live and let live: *"Das ewige gelten lassen, das Leben und leben lassen."* Yet if we live ever so little in conformity with our own moral nature, if we do not fall into the contamination of vanity or the dissociation of pride, then to allow the present moment to have its full value, to compose out of spiritual memories a new present that prepares for an even better one, and via the poetry of the past to move toward the truth of the future—this is at once the seed and the fruit of human wisdom. We must envy Goethe his precious ability whereby he could "by the rapidity and diversity of his thought divide the day into millions of parts and make of them a small eternity."

If the two contrary doctrines of eternity in the present both have a purifying effect, they do not work in the same way.

Contamination aims to make us reach our goal. It is, so to speak, precipitate. Very little is needed to transform it, and indeed history shows that it can assume diverse forms. The pleasure-seeker is on occasion changed into a hero—more often into a saint—and this can only be because some analogy exists between the two natures he assumes, for it is quite impossible to become what one is not. Furthermore, is the Epicurean not more of a man than the impassive Stoic? We come to see that true wisdom is opposed by both false wisdom and folly, just as true peace is opposed by war and false peace, as friendliness is opposed by hostility and false friendliness, just as true religion is opposed by nonbelief and false religion. If we had to choose, in each instance we would choose the inverse forms over the false, because the former contain an image of truth and make conversion possible, whereas, acting behind a mask of truth, the false corrupts essence.

Contamination would not separate. Dissociation has raised an instant of duration above the plane of ordinary time. What an admirable method! In its search for absolute timelessness, humanity has been put on the trail of its most precious acquisitions. Aristotle's eagerness to discover the circular or apodictic reasonings that make pure thought completely independent and convertible led him to formal logic. Plotinus's eagerness to prepare for and to prolong ecstasy, and especially to make it serve in interpreting the universe, led him to his method of ascendant and divisive thought. The same impulse liberated St. Augustine to become the guide of medieval philosophy and mysticism, and his influence is far from being exhausted even for us today. The exercises of mysticism and the methods of prayer have often been motivated by the idea of at least preparing for, if not attaining to, a state of exaltation. Spinoza raised knowledge of the third order above the other two; he breathed spirituality into the domain of mathematical realities—which seemingly would be the most alien of all—and thereby unquestionably gave aid and support to future seekers after truth. Long before him, Pythagoras had understood that a mystique was needed even for numbers. The same is true of sensory dissociation.

Damaging as it is to ordinary morality, and so hard to keep in balance, dissociation has always favored art. Nothing is more enabling for the artist than to separate the "I" into an idea that represents it and a memory that recalls it—a double image that is seen simultaneously. So I do not hesitate to say that doctrines of dissociation, thanks to exercises that they presuppose, favor an initial development of human abilities. In the melange of what we call life, they prevent our worshiping flux, as a confused impulse urges us. Such doctrines are oriented toward ascetisicm, and even when they counsel us to descend the mountain, it is only by virtue of having climbed it that we may do so.

The moral man, I have said, tends to separate *via* and *veritas*: he believes that the path of sacrifice is a means of access to his goal, but that it does not share in the value of the goal. The pantheistic view corrects him. It shows, or rather reminds, him that *via* is already *veritas,* and that the means whereby we hasten on toward eternity are also those by which eternity is communicated to us. Dissociation is the victim of a contrary precipitancy: its error is to confuse *via* with *veritas,* as if the means could be the end, as if there were no life except in the operations of the mind.

Contamination, then, is the image of the goal; dissociation is the image of the method. Time in human life is unlike what either of them assumes. Contamination places man below, dissociation places him above, the level on which his days unfold. Perhaps one must pass through both, as through trial by fire. The further thought progresses, the more common ideas, by virtue of being discovered and rediscovered, will gain in refulgence and inner meaningfulness. They will receive a kind of light reflected from the very systems of thought that have been tested in man's effort to free himself from the burden of their evidence.

The vice of the pseudo eternity philosophers is not that they discuss eternity in the context of time; it is that they exclude all eternity that is beyond time. Whether it contaminates or dissociates, their concept assumes that virtual eternity, above which the passage of time is suspended, deserves to be called eternity if the mind or senses can invest it with actuality. But how is one to assimilate the timelessness of a connection with the concrete eternity of the subject and person? This would mean replacing a reasonable and even necessary mystery with an absolute enigma that grows from an extreme confusion between thought and being. To think eternity and to live eternally can be construed as one and the same thing only if we really insist on deluding ourselves. If we refuse to recognize the structure of time, we have no alternative but to close our eyes to the human condition or to shuttle continually between exaltation and disappointment in the hope that one will correct the other.

I have not wanted to carry our discussion beyond the limits of observation and experience, but I must emphasize that the conclusions we have reached leave some unfilled gaps.

Time, it seems, belongs to a type of potential reality with which nature is constantly confronting us, but that we find difficult to express appropriately. By preference, the human mind fastens on what is stable, complete, or perfect. When we want to designate ways of being that do not possess any such completeness, we lack adequate terminology. Our language reduces time to a mere unraveling of a reality that is rolled back upon itself, or it focuses on change and neglects to notice— or at least neglects to express—the fact that being in motion can maintain identity with itself. Romantic influences have tended to obscure these formulations, which quickly lead us to concepts that are more seductive than enlightening.

We know how Bergson's philosophy suffered from the unhappy exigencies of language: we find, on the one hand, vital time in which everything evolves, but also spiritual time in which being can be corrupted but can also maintain itself or develop; we have now flow, now maturation. Does Aristotle come to the rescue, perhaps, with his idea of potentiality? Aside from the fact that he never completely distinguished potential from logical possibility, Aristotle did say that the potential was only a preparatory phase which had no purpose other than to develop and flower and to wither away in the process. The musician in action retains nothing of the potential musician: form comes to play its role and by virtue of filling everything, replaces everything. Form directs development or, rather, the unrolling of successive phases; it completes succession and replaces it. This is why form may seem to come from without.

Nevertheless, I see no radical conflict between the ideas of Aristotle and Bergson, and the latter's criticisms of the former are not too startling. (Indeed, in my country, where philosophical thought achieved self-awareness by opposing the Aristotelianism of the Schools, such

134

criticism is practically the rule.) It is even possible that some day their ideas will be reconciled. Perhaps we will come to see that, allowing for their very different mentalities and languages, which correspond in part to the relative development of the sciences as each knew them, both men sought to focus attention on the same point. Both were opposed to concepts that declare change is unthinkable. Zeno was their earliest and their shared adversary. But there is no analogy in their methods or language or turn of mind, and from this we can learn something. If we were able to remove ourselves in thought to a period sufficiently distant from and unlike our own, so that the differences between the two philosophers were to seem negligible, then we could say that Aristotle and Bergson were two biologists. One emphasized the *logos,* the other the *bios.* One spoke the language of logic, which is always virtually dissociative, which draws distinctions only to unite. The other spoke the language of life, which necessarily tends to contaminate ideas despite all efforts to purify that language.

Perhaps it is impossible to find a mode of philosophical discourse that can translate without distortion the development of a spiritual reality. In any event, we can take cognizance of the divergence between language and subject, so that in our thinking we can constantly correct or compensate for it. It matters very little, after all, what processes of projection and construction we use, provided we recognize them for what they are. The best thing to do is to select several simultaneously. In this way, the difference in language and mentality that exists can be seen in the identity or analogy of the results, and so becomes valuable to the intelligence. The difference assures us that we have not mistaken the direction of our method with the order of things as they are.

This, it seems to me, is how spiritual life is illuminated by a pure and new light when we translate it at once into the language of logic and the language of life, and so perceive first, where the double symbol converges and second, where it is imperfect. Then, via an inner rectification which is the basis of criticism, we find the essence of time. To be grasped in its full truth, human time must be detached from both eternal and vital time. Temporal development is not the substitution of form for matter, or even the invasion of action into the realm of the potential; no more is it a natural and, as it were, a necessary flowering of life, after the fashion of the human seed. If we were obliged to speak the language of forms, we would define temporal development as a laborious, endlessly precarious inquiry, rather like the individual's effort to develop himself. Speaking the language of life, we would say that the individual's proper work is to accept and to purify his initial *élan* and instinct. To develop is to win possession of the self in the course of a series of circumstances that are intrinsically neutral in that they can equally well nourish or corrupt. (Whence the necessary function of freedom.) In a word, time is the place of spiritual growth. To become what one is, to be what one has— or better still, to have what one is: this is the problem put to us all.

The significance of time is that it allows us to resolve this problem.

Therefore time is necessary for the edification of an eternal, spiritual person. The attribute of time is flow, its essence conservation. Biological time is merely the image of historical time; its beginning, its prime, and its decline have no subsistence and so their final office is to act as symbols. In the animal order, individuals are the servants of life, which uses them to maintain its form. This is why the moment between fecundation and corruption tends to grow shorter, at least for the male who, even if he grows old, does no more than outlive his acts of love. It is quite otherwise in the spiritual order, where the word "life" points to something altogether unlike that self-enclosed movement and obscure drive. The temporal spiritual life is the conservation of what has been in what is, which is necessary if one is to move forward toward what will be. But since, as we have seen, what will be, in the last analysis, is eternity, the future being only its image, then time must be conceived as the preparation for a life of a third order.

The biological rhythm is not abolished in the spiritual life; no more can consciousness and memory, however sublimated one may imagine them, disappear in this third form of existence. I will say, further, that if one strips time of any notion of flight or of *alea,* either contingency may still continue or persist, for nothing forbids us to conceive of a pure and substantial progress that eternally adds good to good and better to better.

The supreme office of time is to prepare for each conscious being organs of vision and of life that cannot mature in our current existence but that will immediately begin to function so soon as a propitious sphere of operation is offered them, just as the organs of the embryo wait throughout their uterine existence for effort, space, and mobility in order to become active.

Temporal reality—which goes by the name of history and memory, of trial and merit—comprises the stuff of these spiritual organs of eternity that have no function in what we call life. We have no words to denote them, any more than an embryo endowed with intelligence and consciousness could name or even conceive of the purpose of the still unformed apparatus that is silently modifying him without being immediately useful to him, but that later will be his tools of communication and equilibrium.

Someone may object that no tool will be needed to adapt to eternity and that any intervention will be useless and therefore should be rejected. But without an analogue for what biology calls the body—or at least without the subsistence of our consciousness and consequently without the persistence of what has been in time—how could the finite being offer that unresisting opposition to the infinite being which it needs in order to enjoy the presence of the infinite without being absorbed in it?

Temporal maturation continues until the moment the spirit disengages itself from the physical being and the social corpus, the double material through which it has come to know itself in the cosmos. At that point, it allows body and personality to take leave of it, so to speak; they were the two envelopes on which the spirit was able to imprint its character but which were not it. This is the moment we call death. This passing on to another mode of existence, which we are unable to imagine, must correspond in consciousness to the feeling of being absolutely present to the self. When at the same moment the biological and social being escapes from the spiritual persona, the latter becomes eternal. Contrary to what occurs in contamination, the flow of eternity, which was impotent in the temporal life, must absorb succession. At last dissociation functions. It is then that time disappears or, rather, that it is fulfilled.